PLANT BREEDING

Related Legislations

The Author

Phundan Singh [b. 1946] hails from a reputed agricultural family of Western Uttar Pradesh [Village-Sakauti, Tehsil-Mawana, District-Meerut]. He graduated in Agriculture from Agra University, Agra and obtained his Master and Doctorate degrees from Kanpur University, Kanpur. He has throughout a brilliant academic record. His area of specialization is Plant Breeding and Genetics.

Dr. P. Singh has total experience of 44 years in the field of Genetics and Plant Breeding. He has served the Central Institute for Cotton Research, Nagpur for about 32 years in various capacities such as Scientist, Senior Scientist, Principal Scientist, Head of Division and Director. He has 300 publications to his credit. He has participated and presented 80 research papers in various National and International Seminars, Symposia and Conferences. He has authored 45 books on Genetics and Plant Breeding mostly published by Kalyani Publishers, New Delhi. He has also authored 18 technical bulletins on various aspects of cotton and contributed 13 chapters in various books.

Dr P. Singh is a member of several Scientific Societies, referee of different journals, Expert member of Selection Committees and M.Sc. and Ph.D. Examiner in different Agricultural Universities. The Institute was recognized as the referral Laboratory for detection of *Bt* gene and received the best Annual Report Award when he was the Director. He has visited several countries such as Russia, Belarus, Ukraine, Crimea, Uzbekistan, Canada, USA, England, Austria and Germany.

PLANT BREEDING

Related Legislations

by

Phundan Singh

Former Director
Central Institute for Cotton Research,
Nagpur – 440 010
Maharashtra, India

2016

Daya Publishing House®

A Division of

Astral International Pvt. Ltd.

New Delhi – 110 002

Cataloging in Publication Data—DK
Courtesy: D.K. Agencies (P) Ltd. <docinfo@dkagencies.com>

Singh, Phundan, 1946- author.
Plant breeding : related legislations / by Phundan Singh.
pages cm
Includes bibliographical references (pages).

ISBN 978-93-5130-854-6 (International Edition)

1. Plant breeding—Law and legislation—India. I. Title.

DDC 346.54046953 23

Published by : **Daya Publishing House®**
A Division of
Astral International Pvt. Ltd.
– ISO 9001:2008 Certified Company –
4760-61/23, Ansari Road, Darya Ganj
New Delhi-110 002
Ph. 011-43549197, 23278134
E-mail: info@astralint.com
Website: www.astralint.com

Laser Typesetting : **Classic Computer Services,** Delhi - 110 035

Printed at : **Thomson Press India Limited**

— Dedicated to —

Dr. Saksham Singh [Daughter in-law]
Er. Rajeev Singh [Son]
Ms. Pavitra Singh [Daughter]

Preface

There are various legislations related to plant breeding and seed technology. Such legislations can be broadly classified into four groups, *viz.* seed legislations, varieties legislations, biodiversity legislations, bio-safety legislations, etc. However, the information on all these legislations is not available from any single source and students have to go through voluminous literature to cover their course requirements. Present book has been designed to provide comprehensive information about all these legislation in one compact volume.

This book covers all plant breeding related legislations in its 20 chapters. The book has been divided in to five sections. Section I deals with seed related legislations consisting of 9 chapters. Section I deals with varieties related legislations in 3 chapters, Section III deals with biodiversity legislations in 2 chapters, Section IV deals with bio-safety legislations in 2 chapters and Section V deals with miscellaneous topics consisting of 4 chapters. Each chapter has been presented point wise and step by step for easy grasping of students. Glossary of technical terms is presented at the end of each chapter. Some useful references are given at the end for gathering further details.

The information contained in this book has been gathered from various published sources and Internet websites. Attempts have been made to provide latest information even then some valuable information might have been missed.

The cooperation extended by my wife Mrs. Jaswanti Singh during preparation of this book is highly appreciable. Hope this book would be useful to the students, researchers, teachers and seed growers engaged in the field of plant breeding and seed technology. Constructive suggestions of students and teachers are invited for further improvement of this book.

Phundan Singh

Contents

Section III: Biodiversity Related Legislations

Section IV: Bio-safety Related Legislations

Section V: Miscellaneous Topics

Abbreviations

AOSCA: Association of Official Seed Certifying Agencies.
BARI: Bangladesh Agricultural Research Institute.
BMCs: Biodiversity Management Committees.
CAUs: Central Agricultural Universities.
CCI and E: Central Committee for Import and Export.
CICR: Central Institute for Cotton Research.
CVRC: Central Variety Release Committee.
CBD: Convention on Biological Diversity.
COP: Conference of Parties.
DAC: Department of Agriculture and Cooperation.
DARE: Department of Agriculture Research and Education.
DBT: Department of Bio-Technology.
DIA: Designated Inspection Agency.
DSR: Directorate of Seed Research.
DUS: Distinctiveness, Uniformity and Stability.
EDV: Essentially Derived Variety.
FAO: Food and Agriculture Organization.
FRA: Farmers Rights Act.
GA: Gibberellic Acid.
GI: Geographical Indication.

HAU: Haryana Agricultural University.
ICAR: Indian Council of Agricultural Research.
ICRISAT: International Crop Research Institute for Semiarid Tropics.
IPRs: Intellectual Property Rights.
IPGRI: International Plant Genetic Resources Institute.
ISTA: International Seed Testing Association.
NBA: National Biodiversity Authority.
NBPGR: National Bureau of Plant Genetic Resources.
NRPV: National Register of Plant Varieties.
NSC: National Seeds Corporation.
NSP: National Seed Project.
OGL: Open General License.
PBR: Plant Breeders Rights.
PBRA: Plant Breeders Rights Act.
PEG: Poly Ethylene Glycol.
PEQ: Post Entry Quarantine.
PFSO: Plants, Fruits and Seeds Order.
PPA: Plant Protection Advisor.
PRA: Pest Risk Analysis.
PPV and FR: Protection of Plant Varieties and Farmers' Rights.
PVR: Plant Variety Registry.
RCGM: Review Committee on Genetic Manipulation.
SAUs: State Agricultural Universities.
SBB: State Biodiversity Board.
SFCI: State Farm Corporation of India.
SPC: Supplementary Protection Certificate.
SPT: Seed Production Technology.
SSC: State Seed Corporation.
SSCA: State Seed Certification Agency.
SVRC: State Variety Release Committee.
TK: Traditional Knowledge.
UPOV: Union for Protection of Plant Varieties.
UAS: University of Agricultural Sciences.

Section I

Seed Related Legislations

Chapter 1

Introductory Chapter

Introduction

Plant breeding can be defined as a science as well as an art of improving the genetic makeup of plants in relation to their economic use. Recently plant breeding has been described as a technology of developing superior crop plants for various purposes (Riley, 1978). Plant breeding is considered as the current phase of crop evolution (Simmonds, 1979).

Art refers to human imagination, creativity and skill. Plant breeding is an art because selection of superior plants requires human skill, imagination and experience.Science refers to systematic knowledge of a subject gained through human observations and experiments involving certain principles. Plant breeding is a science because development of superior varieties or hybrids involves genetic principles, sequential steps (hybridization, selection, evaluation, multiplication, release etc.) and experimentation. It is an applied branch of genetics which is used for improving and changing the genetic makeup of crop plants.

Technology refers to development of a useful commercial product/service involving scientific principles and human skill. In plant breeding, the ultimate aim is to develop superior varieties or hybrids for commercial cultivation.

Thus plant breeding is an art, a science and a technology of developing genetically superior plants in terms of their economic utility for the mankind. Since plant breeding deals with genetic improvement of crop plants, it is also known as the science of crop improvement.

Areas/Activities of Plant Breeding

There are three important areas or activities of plant breeding, *viz.* (i) varietal development, (ii) seed multiplication and (iii) varietal maintenance (Figure 1.1). These are defined as follows:

1. **Varietal Development:** It makes use of various breeding methods such as selection, hybridization [pedigree, backcross and bulk methods], mutation and genetic engineering to develop new plant varieties.

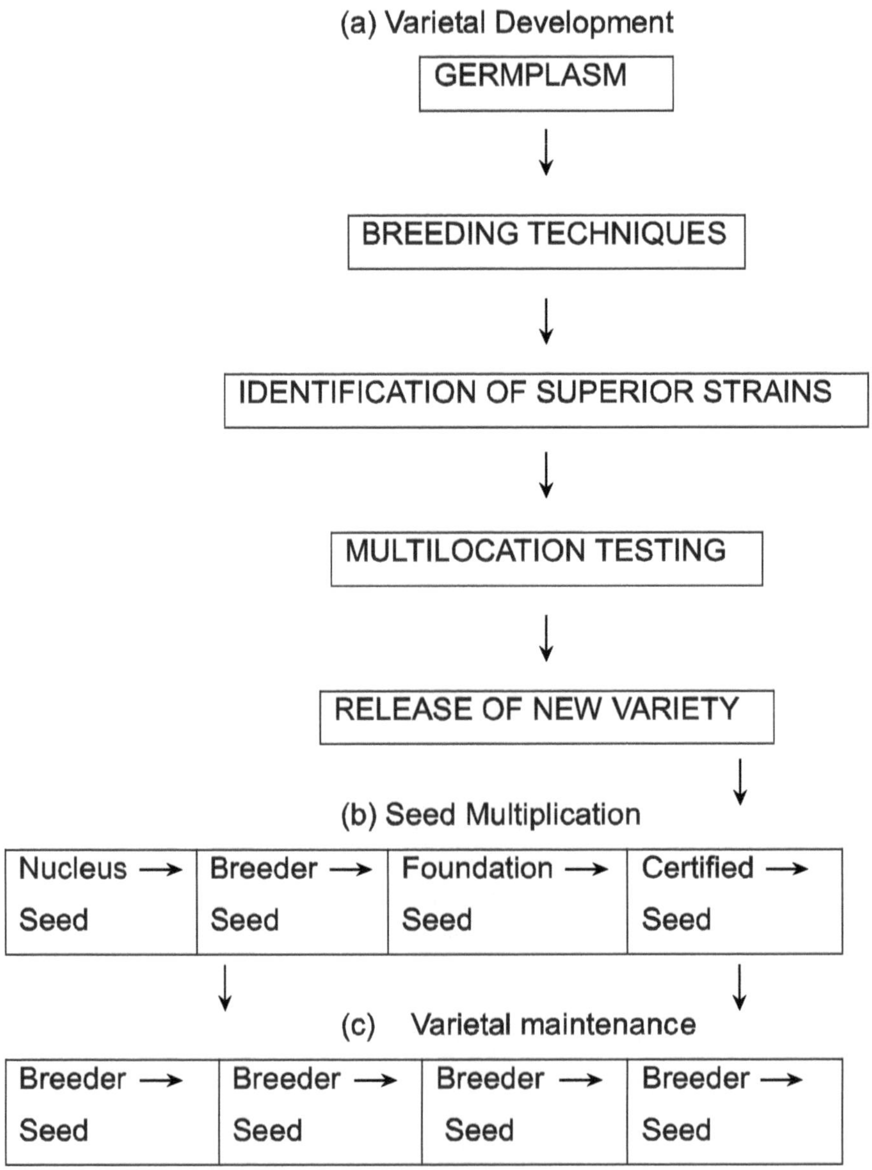

Figure 1.1: Plant Breeding Activities.

2. **Seed Multiplication:** It deals with multiplication or production of successive generations of various categories of seeds such as nucleus seed, breeder seed, foundation seed and certified seed. The seed multiplication or increase is regulated by seed certification agencies and breeders.
3. **Varietal Maintenance:** It deals with continuous fresh supply of breeder seed which is used to start another cycle of seed multiplication. It make use of various selection/maintenance procedures.

Plant Breeding Related Legislations

Government of India has formulated, approved and enacted various legislations related to plant breeding from time to time. Various legislations related to plant breeding and genetics approved and implemented by the Government of India so far can be broadly divided into four major groups, *viz.*, seed legislations, varieties legislations, biodiversity legislations and bio-safety legislations (Table 1.1).

Table 1.1: List of Plant Breeding Related Legislations

Sl.No.	*Areas of Legislations*	*List of Legislations*
1.	Seed Related Legislations	Seed Act, 1966
		Seed Rules, 1968
		Seed Amendment Rules, 1972
		Seed Amendment Rules, 1973
		Seed Amendment Rules, 1974
		Seed Amendment Rules, 1981
		New Seed Policy, 1988
		Plants, Fruits and Seed Orders, 1989
		National Seed Policy, 2002
		Seed Act, 2004
2.	Varieties Related Legislations	Union for Protection of Plant Varieties (UPOV) Act, 1978
		UPOV Act 1991
		Protection of Plant Varieties and Farmers Rights, 2001
		Protection of Plant Varieties Rules, 2003
		Farmers' Rights Act 2001
		Plant Breeders' Rights Act
3.	Biodiversity Related Legislations	Convention on Biodiversity
		Indian Biodiversity Legislation
4.	Bio-safety Related Legislations	Cartagena Protocol on Bio-safety
		Bio-safety Regulations for Transgenic Plants
5.	Miscellaneous Legislations	Plant Quarantine Order, 2003
		Intellectual Property Rights.

1. **Seed Related Legislations:** In India, the Seed Act was first introduced in 1966 which is called as Seeds Act 1966. This Act was amended in 1972 which is known as the Seeds amended Act 1972. Subsequently, the seed act was thoroughly revised and the revised act was approved by the Government of India in September 2004. The revised Seeds Act is known as Seeds Act 2004 which came into force with effect from January 2005. It replaces the Seeds Act, 1966. The Indian seed Act has undergone several changes over the years. Seed legislations were formulated and implemented with the following objectives.
   (i) To ensure availability of quality seeds to the farmers.
   (ii) To regulate production and marketing of planting seed through various government organizations such as national Seeds Corporation, State Seeds Corporations, State Seed Certification Agencies, State Agricultural Universities and Central Agricultural Universities.
2. **Varieties Related Legislations:** Various plant breeding legislations related to plant varieties have been formulated and implemented both at International and national levels. International legislations include Union for Protection of Plant Varieties such as UPOV Act, 1978 and UPOV Act, 1991. In India, Protection of Plant Varieties and Farmers' Rights Act was approved by the government of India in 2001. Other legislations related to plant varieties include Plant Breeders Rights and farmers Rights. In India, these have been covered by Protection of Plant Varieties and Farmers' Rights Act. Varieties related legislations were formulated and implemented with the following objectives.
   (i) To provide legal rights to the original plant breeders to regulate production and marketing of his varieties. In other words, original plant breeder can take advantage of varieties developed by him/her.
   (ii) To provide legal rights to the farmers to use protected plant varieties for commercial crop production. He can use seed of his crop for raising further crop or he can exchange seed of such varieties with other farmers.
   (iii) To make use of protected plant varieties by plant breeders for developing new superior crop varieties.
3. **Biodiversity Related Legislations:** The Convention on Biological Diversity is also known as the Biodiversity Convention. It is an international treaty that was adopted in Rio de Janeiro in June 1992. The Convention was opened for signature at the Earth Summit in Rio de Janeiro on 5 June 1992 and entered into force on 29 December 1993. The biodiversity agreement has been signed by 189 countries. In India the Biological Diversity Act [also called biodiversity act] was passed by the Central Government in 2002.Biodiversity related legislations have the following three main goals:
   (i) Conservation of biological diversity (or biodiversity):
   (ii) Sustainable use of its components; and
   (iii) Fair and equitable sharing of benefits arising from genetic resources.

4. **Bio-safety Related Legislations:** The Cartagena Protocol on Bio-safety is the first international agreement to regulate the trans-boundary movements of genetically engineered (GE) organisms. The Bio-safety Protocol is a subsidiary agreement to the UN Convention on Biological Diversity (CBD), which was signed by over 150 governments at the Rio Earth Summit in 1992. The Bio-safety Protocol seeks to protect biological diversity from the potential risks posed by living modified organisms [LMOs] resulting from modern biotechnology.
5. **Miscellaneous Legislations:** In addition to above, there are some other legislations which are related to plant breeding and genetics. Such legislations include Plant Quarantine Order, 2003, Intellectual Property Rights, etc. This section also deals with World Trade Organization and Globalization which are important in the present context.

Brief History

Seed Act refers to the legal procedures approved by the government for production and marketing of seeds. The main objective of the Seeds Act is to ensure availability of quality seeds to farmers. In India, the Seed Act was first introduced in 1966 which is called as Seeds Act 1966. This Act was amended in 1972 which is known as the Seeds amended Act 1972. Subsequently, the seed act was thoroughly revised and the revised act was approved by the Government of India in September 2004. The revised Seeds Act is known as Seeds Act 2004 which came into force with effect from January 2005. It replaces the Seeds Act, 1966. The Indian Seed Act has undergone several changes over the years. Important events related to plant breeding related legislations are presented in Table 1.2.

Table 1.2: Important Events related to Plant Breeding Legislations in India

Year	*Important Event*	*Remarks/Brief Description*
1966	The first Indian Seed Act was formulated	To provide good quality planting seeds to the farmers.
1968	Seed Rules were framed to implement various legislations given Seed Act 1966.	Enactment of Seed Act 1966.
1972	The Seed Act 1966 was amended known as seed amendment Rules 2002.	The Jute seeds were included to the Seeds Act, establishment of Seed Certification Board and fixing of minimum standards.
1973	The Seed Act 1966 was amended known as seed amendment Rules 2003.	Power of appellate authority and duty of seed analyst were slightly modified. The Seed Testing Manual was published by ICAR for reference.
1974	The Seeds [amendment] Rules 1974 were introduced.	This amendment conferred more powers on seed inspectors during crop failure.
1978	The UPOV Seed Act was introduced	The new variety should have distinctiveness, uniformity and stability for protection under this act.

Contd...

Table 1.2–*Contd...*

Year	*Important Event*	*Remarks/Brief Description*
1981	The Seeds [amendment] Rules 1981 were introduced.	Adoption of Minimum Indian Seed Certification Standards published by Central Seed Committee was made compulsory for seed certification.
1983	The Seeds [control] Order 1983 was introduced.	This Order included Seeds as essential commodity item under the Commodity Act 1955.
1988	The New Policy on Seed Development was introduced.	The new policy was formulated to provide Indian Farmers with access to the best available seeds and planting materials of domestic as well as exotic origin.
1989	Plants, Fruits and Seeds Order 1989 was introduced.	This replaces the Plants, Fruits and Seeds Order 1984 and provides regulations for post entry quarantine checks.
1991	The UPOV Seed Act 1991 was introduced.	The new variety should have novelty in addition to distinctiveness, uniformity and stability for protection under this modified UPOV Act.
2001	The Protection of Plant varieties and Farmers Rights Act was formulated.	A new variety with novelty, distinctiveness, uniformity and stability can be registered for protection under this Act.
2002	The National Seed Policy was formulated.	The National Seed Policy was formulated to raise Indias' share in the global seed trade by facilitating advanced scientific aspects such as biotechnology to farmers and in March 2002 first transgenic cotton was approved for commercial cultivation in India.
2003	Plant Quarantine [Regulation of Import into India] was introduced.	This Order has now replaced the Plants, Fruits and Seeds Order 1989.
2003	Protection of Plant Varieties Rules 2003 were introduced.	These rules were introduced for smooth implementation of the PPV and FR Act 2001.
2004	The New Seed Bill was formulated and introduced.	This Seed Bill has replaced the Seeds Act 1966. Registration of all seed materials is compulsory and self certification is permitted.

Summary

Plant breeding can be defined as a science as well as an art of improving the genetic makeup of plants in relation to their economic use. Recently plant breeding has been described as a technology of developing superior crop plants for various purposes (Riley, 1978). Plant breeding is considered as the current phase of crop evolution (Simmonds, 1979). There are three important areas or activities of plant breeding, *viz.* (i) varietal development, (ii) seed multiplication and (iii) varietal maintenance.

Government of India has formulated, approved and enacted various legislations related to plant breeding from time to time. Various legislations related to plant breeding and genetics approved and implemented by the Government of India so far can be broadly divided into four major groups, *viz.*, seed legislations, varieties legislations, biodiversity legislations and bio-safety legislations.

Seed Act refers to the legal procedures approved by the government for production and marketing of seeds. The main objective of the Seeds Act is to ensure availability of quality seeds to farmers. In India, the Seed Act was first introduced in 1966 which is called as Seeds Act 1966. This Act was amended in 1972 which is known as the Seeds amended Act 1972. Subsequently, the seed act was thoroughly revised and the revised act was approved by the Government of India in September 2004. The revised Seeds Act is known as Seeds Act 2004 which came into force with effect from January 2005. It replaces the Seeds Act, 1966. The Indian seed Act has undergone several changes over the years.

Questions

1. **Define plant breeding and describe important areas of plant breeding.**
2. **List important plant breeding related legislations. Describe objective of any three legislations.**
3. **Describe the objectives of the following plant breeding related legislations.**
   (i) Seed Act, 1966 **(ii)** Seed Act, 2004
   (iii) Seed Rules, 1968 **(iv)** National Seed Policy, 2002
4. **Describe the objectives of the following plant breeding related legislations.**
   (i) UPOV Act 1978 **(ii)** UPOV Act, 1991
   (ii) Convention on Biodiversity **(iv)** Cartagena Protocol
5. **Write short notes on the following:**
   (i) New Seed Policy, 1988 **(ii)** PFS Orders, 1989
   (ii) Seed Amendment Rules, 1972 **(iv)** Seed Amendment Rules, 1981
6. **Write short notes on the following:**
   (i) Seed Amendment Rules, 1973 **(ii)** Seed Amendment Rules, 1974
   (iii) PPV Rules, 2003 **(iv)** Seed Control Order, 1983
7. **Define the following terms:**
   (i) Biodiversity **(ii)** Bio-safety
   (ii) Seed legislations **(iv)** Farmers Rights

Chapter 2

Indian Seed Act 1966

Introduction

Seed Act refers to the legal procedures approved by the government for production and marketing of seeds. The main objective of the Seeds Act is to ensure availability of quality seeds to farmers. In India, the Seed Act was first introduced in 1966 which is called as Seeds Act 1966. This Act was amended in 1972 which is known as the Seeds amended Act 1972.

Salient Features

The main features of Indian Seed Act 1966 in relation to coverage, seed category, license, registration, criteria for protection, period of protection, certification, compensation, farmers' privilege, penalty, and constitution of central seed committee are briefly presented below.

1. **Coverage:** This Act has narrow coverage than seeds act of 2004. It includes agriculture and horticulture only.
2. **Seed Category:** This act permits sale of certified seed only The name of variety, physical purity and germination percentage have to be indicated on the seed container or bag.
3. **License:** The seed dealers, sellers and growers should have license which is compulsory.
4. **Registration:** Registration for all kinds and varieties of seed is compulsory. No producer can grow seeds unless he is registered. It is compulsory for all cultivated varieties of field crops, vegetable crops and fruit crops. Every seed producer and dealer, and horticulture nursery has to be registered with the state government.

5. **Criteria for Registration:** The released and notified varieties can be registered for seed multiplication.
6. **Period of Registration:** In this seed Act the period of registration has not been defined.
7. **Certification:** The Act does not permits self certification of seeds by accredited agencies. However, it allows the central government to recognize certification by foreign seed certification agencies. Seed producers are not permitted to self certify the performance of their seed under certain conditions.
8. **Compensation:** The disputes about the quality of seeds have to be settled in the consumer court. Any loss in the production due to poor seed quality is claimed in the consumer court.
9. **Farmers' Privilege:** This Act allows farmers' privilege. The Bill does not restrict the farmer's right to use or sell his farm seeds and planting material, provided he does not sell them under a brand name.
10. **Penalty:** Any person who contravenes any provisions of the Act or imports, sells or stocks seeds deemed to be misbranded or not registered, can be punishable by a fine between Rs 100-1000/and 6 months imprisonment.
11. **Central Seed Committee:** In this Act, the central seed committee consists of Chairman [Secretary DARE], six persons related to interest in quality seeds, two representatives of seed growers and one representative from each state. Members are nominated for a period of two years and can be re-nominated for a similar period.

Merits of Seed Act, 1966

There are several merits of seed act 1966. Merits of this act related to biodiversity, seed monopoly, tracing defective seed lot, certification, farmers rights, seed quality, duration of registration and effects on seed industry and productivity are briefly presented as follows.

1. **Effect on Biodiversity:** This seed act permits cultivation of both new varieties as well as land races. Thus it does not t have any adverse effect on crop biodiversity. Thus there is no Thus there is no danger of uniformity.
2. **No Monopoly of Seed**: When the seed act 1966 was approved, there were no multinational seed companies as a result there was no risk of monopoly of seed. The entire seed production was under government organizations.
3. **Easy to Trace Defective Seed Lot:** The Seed Act does provides for a mechanism to trace back a packet of seed to the dealer, processor and producer. Thus tracing of defective lot was easy to rectify any deficiencies in the supply chain.
4. **No Self Certification**: There was no provision of self certification in the Seed Act 1966. The certification is done by government seed certification agencies. Thus it does not have any scope for false certification of the seed.

5. **Permits Farmers' Rights:** It permits farmers' rights. The farmers need not to register their varieties.
6. **Ensures Quality Seed:** This act ensures availability of good quality seeds to farmers by imposing penalties on the sale of spurious seeds.
7. **Registration for long Duration:** This act provides registration for a long period. Because the period of registration has not been specified.
8. **Promotes Seed Industry:** The introduction of Seed Act, 1966 resulted in the fast development of seed industry in India due to involvement of representatives of multinational seed companies or industries in the central seed committee.
9. **Increase in Productivity:** The Seed act has helped in increasing crop productivity due to availability of better quality seeds.

Table 2.1: Merits and Demerits of Seed Act 1966

Sl.No.	*Particulars*	*Comments or Explanation*
A.	**Merits**	
1.	Farmers privilege	Farmers' rights or privilege are protected.
2.	Effect on biodiversity	There is no adverse effect on biodiversity because this act permits cultivation of both improved varieties and land races.
3.	Certification	Certification is permitted by government organizations only which ensures quality of seed.
4.	Tracing defective seed lot.	The tracing of defective seed lot is easy.
5.	Registration	It permits registration till the variety is in seed production chain.
6.	Licensing	Licensing is compulsory for seed sellers which ensures seed quality.
B.	**Demerits**	
1.	Coverage	Coverage is narrow. It includes Agriculture and horticulture only.
2.	Central Seed Committee	Central Seed Committee does not have representative of private seed industry.
3.	Compensation	Not clearly defined.
4.	Seed Sale	Not clearly defined
5.	Benefit Sharing	There is no provision of benefit sharing.
6.	Period of protection	Not defined.

Demerits of Seed Act 1966

There are some demerits of Seed Act 1966. Demerits related to coverage, constitution of central seed committee, etc. are presented below.

1. **Narrow Coverage:** This act has narrow coverage [agriculture and horticulture] than seed act 2004 which covers, forestry, plantation crops, medicinal and aromatic plants, in addition to agriculture and horticulture.

2. **Central Seed Committee:** In the Seed Act 1966, the size of central seed committee is bigger than that of Seed Act 2004. There was no representative from the private seed industry.
3. **Compensation:** It is not clear whether the compensation would include the value of the crop or only the cost of the seed.
4. **Seed Sale:** It is not clear whether a seed producer may sell seed which is registered by a different producer.
5. **Benefit Sharing:** The Act does not have the provision of benefit sharing unlike the Convention on Biological Diversity and the PPVFR Act.

Summary

Seed Act refers to the legal procedures approved by the government for production and marketing of seeds. The main objective of the Seeds Act is to ensure availability of quality seeds to farmers. In India, the Seed Act was first introduced in 1966 which is called as Seeds Act 1966. The main features of Indian Seed Act 1966 in relation to coverage, seed category, license, registration, criteria for protection, period of protection, certification, compensation, farmers' privilege, penalty, exclusion, research exemption and constitution of central seed committee are briefly presented.

There are several merits of seed act 1966. Merits of this act related to biodiversity, seed monopoly, tracing defective seed lot, certification, farmers rights, seed quality, duration of registration and effects on seed industry and productivity are briefly presented.

There are some demerits of Seed Act 1966. Demerits related to coverage, constitution of central seed committee, etc. are presented.

Questions

1. **Explain in brief main features of Indian Seeds Act 1966.**
2. **What are merits and demerits of Indian Seeds Act 1966?**
3. **Give a comparison of Indian Seeds Act 1966and Seed Act 2004in a tabular form.**
4. **Describe the constitution of Central Seed Committee of**
   (i) Indian Seeds Act 1966 **(ii)** Indian Seeds Act 2004.
5. **What do the following terms signify?**
   (i) Extant Variety **(ii)** Transgenic variety
   (iii) Spurious Seed **(iv)** Farmers Rights
6. **Write short notes on the following:**
   (i) Farmers' Rights **(ii)** Farmers' Compensation
   (iii) Orthodox seeds **(iv)** Recalcitrant seeds

7. **Describe briefly India Seeds Act 1966 on the basis of following:**
   (i) Seed certification
   (ii) Registration
   (iii) Licensing
   (iv) Penalties

8. **Describe briefly the Structure of Central Seed Committee of Indian Seed Act 1966.**

9. **What are the similarities and dissimilarities between Seeds Act 1966 and Seeds Act 2004?**

10. **Write Short Notes on the following:**
   (i) Indian Seeds Act 1966
   (ii) Indian Seeds Act 2004
   (iii) PPV and FR Act 2001
   (iv) Biodiversity Act 2002

Chapter 3

The Seed Rules, 1968

Introduction

The Seeds Rules, 1968 under Seed Act, 1966 (Act No. 54 of 1966) were formulated and approved for implementation of Seed Act, 1966.These rules consist of 11 Parts and 39 sections.

Salient Features

Salient features of seed rules, 1968 related to scope and coverage, central seed committee, central seed laboratory, seed certification agencies, marking and labeling, requirements, certification of seeds, appeals, qualification and duties of seed analysts and seed inspectors, sealing, fastening, dispatch and analysis of samples and miscellaneous matters are briefly presented as follows.

1. Scope and Coverage

These rules may be called the Seeds Rules, 1968. and extend to whole of India. These rules cover agriculture and horticulture. Various terms used related to seed rules have also been defined [Part 1].

2. Central Seed Committee

This committee has several functions. Main functions of this committee are [Part 2]:

(i) To recommend the rate of fees to be levied for analysis of samples by the Central and States Seed Testing Laboratories and for certification by the certification agencies;

(ii) To advice the Central or State Governments on the suitability of seed testing laboratories;

(iii) To send its recommendations and other concerning records to the Central Government;

(iv) To recommend the procedure and standards for certification, tests and analysis of seeds; and

(v) To carry out any other functions related to these rules.

3. Central Seed Laboratory

In addition to the functions entrusted to the/central Seed Laboratory by the Act, the Laboratory shall carry out the following functions [Part 3].

(i) Initiate testing programs in collaboration with the State Seed Laboratories designed to promote uniformity in test results between all seed laboratories in India;

(ii) Collect data continually on the quality of seeds found in the market and make this data available to the Committee; and

(iii) Carry out such other functions as may be assigned to it by the Central Government from time to time.

4. Seed Certification Agency

In addition to the functions entrusted to the certification agency by the Act, the Agency shall certify seeds of any notified kinds or varieties [Part 4];

(i) Certify seeds of any notified kinds or variety;

(ii) Outline the procedure of submission of applications and for growing, harvesting, processing, storage and labeling of seeds intended for certification till the end to ensure that seed lots finally approved for certification are true to variety and meet prescribed standards for certification under the Act or these rules;

(iii) Maintain a list of recognized breeders of seeds;

(iv) Verify, upon receipt of an application, for certification, that the variety is eligible for certification that the seed source need for planting was authenticated and the record of purchase is in accordance with these rules and the fees has been paid;

(v) Take sample and inspect seed lots produced under the procedures laid down by the certification agency and have such samples tested to ensure that the seed conforms to the prescribed standards of certification;

(vi) Inspect seed processing plants to see that the admixtures of other kinds and varieties are not introduced;

(vii) Ensure that action at all stages *e.g.* field inspection, seed processing, plant inspection, analysis of samples taken and issue of certificates (including tags, marks, labels and seals) is taken expeditiously;

(viii) Carry out educational programs designed to promote the use of certified seed including a publication listing certified seed growers and sources of certified seed;

(ix) Grant certificate (including tags, marks, labels and seals etc.) in accordance with the provisions of the Act and these Rules;

(x) Maintain such records as may be necessary to verify that seed plants for the production of certified seed were eligible for such planting under these rules;

(xi) Inspect fields to ensure that the minimum standards for isolation, rouging (where applicable) use of male sterility (where applicable) and similar factors are maintained at all times, as well as ensure that seed borne diseases are not present in the field to a greater than those provided in the standards for certification.

5. Marking or Labeling

Marketing and labeling decides about various aspects such as labeling, contents of label and procedure of labeling as discussed below [Part 5].

(a) Responsibility for marking or labeling**:** When seed of a notified kind or variety is offered for sale, each container shall be marked or labeled in the manner hereinafter specified. The person whose name appears on the mark or label shall be responsible for the accuracy of the information required to appear on the mark

or label so long as seed is contained in the unopened original container.

Provided, however, that such person shall not be responsible for the accuracy of the statement appearing on the mark or label if the seed is removed from the original unopened container, or he shall not be responsible for the accuracy of the germination statement beyond the date of validity indicated on the mark or label.

(b) **Contents of the mark or label**: There shall be specified on every mark or label

(i) Particulars, as specified by the Central Government under Clause (b) of section 6 of the act.

(ii) A correct statement of the net content in terms of weight and expressed in metric system.

(iii) Date of testing.

(iv) If the seed in container has been treated.

a. Statement indicating that the seed has been treated.

b. The commonly accepted chemical of abbreviated chemical (generic) name of the applied substance; and

c. If the substance of the chemical used for treatment, and present with the seed is harmful to human beings or other vertebrate animals, a caution statement such as "Do not use for food, feed or oil purposes". The caution for mercurial and similarly toxic substance shall be the word "Poison" which shall be in type size, prominently displayed on the label in red.

(v) The name and address of the person who offers for sale, sells or otherwise supplies the seed and who is responsible for its quality;

(vi) The name of the seed as notified under section 5 of the act.

(c) Manner of marking or labeling the container under clause (c) of section 7 and clause (b) of section 17 :

(i) The mark or label containing the particulars of the seed shall appear on each container of seed or on a tag or mark or label attached to the container in a conspicuous place on the inner most container in which the seed is packed and on every other covering in which that container is packed and shall be legible.

(ii) Any transparent cover or any wrapper, case or other covering used solely for the purpose of packing of transport or delivery need not be marked or labeled.

(iii) Particulars that are required to be displayed on a label on the container, be attached, painted or otherwise indelibly marked on the container.

(d) Mark or label not to contain false or misleading statement:

The mark or label shall not contain any statement, claim, design, device, fancy name or abbreviation which is false or misleading in any particular concerning the seed contained in the container.

(e) The mark or label shall not contain any reference to the Act, or any of these rules or any comment on, or reference to, or explanation of any particulars or declaration required by the Act or any of these rules which directly or by implication contradicts, qualifies or modifies such particulars or declaration.

(f) Denial of responsibility for mark or label content prohibited:

Nothing shall appear on the mark or label or in any advertisement pertaining to any seed of any notified kind or variety which shall deny responsibility for the statement required by or under the Act to appear on such mark, label or advertisement.

6. Requirements

There are certain conditions that are required for a person to carry out the seed business. Important requirements to carry out seed business are as follows [Part 6].

(i) No person shall sell, keep for sale, offer to sell barter or otherwise supply any seed of any notified kind or variety, after the date recorded on the container, mark or label as the date up to which the seed may be expected to retain the germination not less than that prescribed standard.

(ii) No person shall alter, obliterate or deface any mark or label attached to the container of any seed.

(iii) Every person selling, keeping for sale, offering to sell, bartering or otherwise supplying any seed of notified kind or variety [under Section 7], shall keep over a period of three years a complete record of each lots of seed sold except that any seed sample may be discarded one year after the entire lot represented by such sample has been disposed off. The sample of seed kept as part of the complete record shall be as large as the size notified in the official Gazette.This sample, if required to be tested only for determining the purity.

(iv) Classes and sources of certified seed:

There shall be three classes of certified seed, namely, foundation, registered and certified, and each class shall meet the following standards for that class.

a. **Foundation seed:** This shall be the progeny of breeder's seed, or be produced from foundation seed which can be clearly traced to breeder's seed. Its production shall be supervised and approved by a seed certification agency in such a way to maintain specific genetic purity and meet certification standards prescribed for the crop under certification.

b. **Registered seed:** This shall be the progeny of foundation seed and is produced in such a way to maintain its genetic purity and meet germination standards for the particular crop being certified.

c. **Certified seed:** This shall be the progeny of registered or foundation seed ans is produced in such a way to maintain genetic purity and germination standards specified for the particular crop being certified.

At the discretion of the certification agency (when considered necessary to maintain adequate seed supplies) certified seed may be progeny of certified seed provided this reproduction may not exceed three generations and provided further that it is determined by the seed certification agency that the genetic purity will not be significantly altered.

7. Certification of Seeds

This section deals with application, fees and certificate. These are explained as follows [Part 7].

(i) Application for the grant of a certificate: Every application for the grant of a certificate shall be made in prescribed Form and contain the following particulars,

a. the name, profession and place of residence of the applicant;

b. the name of the seed to be certified, its notified kind or variety;

c. class of the seed;

d. source of the seed;

e. limits of germination and purity of the seed;

f. mark or label of the seed.

(ii) Fees: Every application shall be accompanied by a fee of Rs,25 in cash.

(iii) Certificate: Every certificate granted shall be in Form II and shall be granted by the certification agency after making enquiries and satisfying itself.

I. The person to whom the certificate is granted shall attach a certification tag to every container of the certified seed and shall follow the provisions in respect of marking or labeling provided by or under the act.

II. The certification tag shall contain the following particulars namely.
   a. Name and address of the certification agency.
   b. Kind and variety of the seed.
   c. Lot no. or other mark of the seed.
   d. Name and address of the certified seed producer.
   e. Date of issue of the certificate and of its validity.
   f. An appropriate sign to designate certified seed.
   g. An appropriate word denoting the class designation of the seed.

III. The color of the certification tag shall be white for foundation seed purple for registered seed and blue for certified seed.

IV. The container of the certified seed shall carry a seal of such material and in form as the certification agency may determine and no container carrying a certification tag shall be sold by the person if the tag or seal has either been tampered with or removed.

V. The certification tag on the container shall specify.
   a. The period during which the seed shall be used for sowing or planting.
   b. That the use of seed after the expiry of the validity period by any person is entirely at his risk and the holder of the certificate shall not be responsible for any damage to the buyer of the seed.
   c. That no one should purchase the seed if the seal or the certification tag has been tampered with.

VI. The holder of the certificate shall keep record of the details of each lot of the seed which is issued for sale in such form as to be available for inspection and to be easily identified by reference to the number of the lot as shown in the certification tag of each container and such records shall be retained in the case of a seed for which expiry date is fixed for a period of two years from the expiry of such date.

VII. The holder of the certificate shall allow any Seed Inspector, authorized in writing by the certification agency in that behalf, to enter with or without prior notice, the premises where the seeds are grown, processed and sold and to inspect premises, plant and the process of processing at all reasonable hours.

VIII. The holder of the certificate shall allow the Seed Inspector, authorized in writing by the certification agency, to inspect all registers and records maintained under these rules and to take samples of the seeds and shall supply to the Seed Inspector such information as he may require for the purposes of ascertaining whether the conditions subject tom which the certificate has been granted, have been complied with.

IX. The holder of the certificate shall on request furnish to the certification agency from every lot of the seed or from such lot or lots as the said agency may from time to time specify, a sample of such quantity as the agency may consider adequate for any examination required to be made.

X. If the certification agency so directs, the holder of the certificate shall not sell or offer for sale any lot in respect of which a sample is furnished under the proceeding clause until the agency authorizes the sale of such lot.

XI. The holder the certificate shall, on being directed by the certification agency that any part of a lot has been found by the said agency not to conform to prescribed standards of quality or purity specified by or under the Act, withdraw the remainder of that lot from sale and so far as may, in the particular circumstances of the case, be practicable, recall all issues already made from that lot.

XII. The holder of the certificate shall comply with the provisions of the Act and these Rules and with the directions given after not less than one month's notice by the certification agency to such holder.

The Certification agency shall, before granting the certificate, ensure that the Seed conforms to the standards laid down in the Manual known as "Indian Minimum Seed Certification Standards" published by the Central Seed Committee, as amended from time to time".

8. Appeals

All appeals are made in a prescribed form and a prescribed fee has to be paid as discussed below [Part 8].

(i) Every memorandum of appeal shall be in writing and shall be accompanied by a copy of the decision of the certification agency against which it has been preferred and shall set forth concisely and under distinct heads the grounds of objections to such decision without any argument, or narrative.

(ii) Every such memorandum of Appeal shall be accompanied by a treasure receipt for sum of 100/- rupees.

(iii) Every such memorandum of appeal may be presented either in person or through an agent duly authorized in writing in this behalf by the appellant or may be sent by the registered post.

(iv) In deciding appeals under the Act the Appellate authority shall exercise all the powers which a Court has and shall follow the same procedure with a Court follows in deciding appeals from the decree or order of an original Court under the Code of Civil Procedure, 1908 (5 of 1908).

9 (a). Qualifications and Duties of Seed Analysts

This section deals with qualifications of seed analysts, seed inspectors and their duties [Part 9].

1. **Qualifications of Seed Analysts:** A person shall not be qualified for appointment as Seed Analyst unless he.

i. Possesses a Master's or equivalent degree in Agriculture or Agronomy or Botany or Horticulture of a University recognized for this purpose by the Government and has had not less than one year's experience in seed technology; or

ii. Possessed a Bachelor's degree in Agriculture or Botany of a University recognized for this purpose by the Government and has had not less than three years' experience in seed technology.

2. **Duties of a Seed Analyst:**

(i) On receipt of a sample for analysis the Seed Analyst shall first ascertain that the mark and the seal are intact.

(ii) The Seed analyst shall analyze the samples in accordance with the procedures laid down in the Seed Testing Manual published by the Indian Council of Agriculture Research.

(iii) The Seed Analyst shall deliver in Form VII, a copy of report of the result of the seed not later than 30 days from the date of receipt of samples sent by the Seed Inspector.

(iv) The Seed Analyst shall from time to time forward to the State Government the reports giving the result of analytical work done by him.

9. (b) Qualification and Duties of Seed Inspectors:

1. **Qualifications:** A person shall not be qualified for appointment as Seed Inspector unless he is a graduate in Agriculture of a University recognized for the purpose by the Government and has had not less than one year's experience in seed production, or seed development in seed analysis or testing in seed testing laboratory.

2. **Duties of a Seed Inspector:** In addition to the duties specified by the Act, the Seed Inspector shall a. Inspect as frequently as may be required by certification agency all places used for growing storage or sale of any seed of any notified kind or variety;

(i) Satisfy himself that the conditions of the certificates are being observed;

(ii) Procedure and send for analysis, if necessary, samples of any seeds, which he has reason to suspect are being produced, stocked or sold or exhibited for sale on contravention of the provisions of the Act or these rules;

(iii) Investigate any complaint, which may be made to him in writing in respect of any contravention of the provisions of the Act or these rules.

(iv) Maintain a record of all inspections made and action taken by him in the performance of his duties including the taking of samples and the seizure of stocks and submit copies of such record to the Director of Agriculture or the certification agency as may be directed in this behalf.

(v) When so authorized by the State Government detain imported containers which he has reason to suspect contain seeds, import of which is prohibited except and in accordance with the provisions of the Act or these rules.

(vi) Institute prosecutions in respect of breaches of the Act or these rules.

(vii) Perform such other duties as may be entrusted to him by the State Government.

Action to be taken by the Seed Inspector if a complaint is lodged with him:

1. If farmer has lodged a complaint in writing that the failure of the crop is due to the defective quality of seeds of any notified kind or variety supplied to him, the Seed Inspector shall take in his possession the marks or labels, the seed containers and a sample of unused seeds to the extent possible from the complaint for establishing the sources of supply of seeds and shall investigate the causes of the failure of his crop by sending samples of the lot to the Seed Analyst for detailed analysis at the State Seed Testing Laboratory. He shall thereupon submit the report of his findings as soon as possible to the competent authority.
2. In case, the Seed Inspector comes to the conclusion that the failure of the crop is due to the quality of seeds supplied to the farmer being less than the minimum standards notified by the Central Government, he shall launch proceedings against the supplier for contravention of the provisions of the Act or these Rules

10. Sealing, Fastening, Dispatch and Analysis of Samples

This section deals with procedure of sampling, packing, fastening and sealing samples, containers to be used, labeling, form of seed report, fees, etc. [Part 10]. These aspects are briefly explained as follows.

(i) Manner of taking samples: Samples of any seed of any notified kind of variety for the purpose of analysis shall be taken in a clean dry container which shall be closed sufficiently tight to prevent leakage and entrance of moisture and shall be carefully sealed.

(ii) Containers to be labeled and addressed: All containers containing samples for analysis shall be properly labeled and the parcels shall be properly addressed. The label on any sample of seed sent for analysis shall bear.

a. Serial number;
b. Name of the sender with official designation, if any;
c. Name of the person from whom the sample has been taken.
d. Date and place of taking the sample;
e. Kind or variety of the seed for analysis;
f. Nature and quantity of preservative, if any, added to the sample.

(iii) Manner of packing, fastening and sealing the samples: Samples of seed sent for analysis shall be packed, fastened and sealed in the following manner.

a. The stopper shall first be securely fastened to as to prevent leakage of the containers in transit;

b. The container shall then be completely wrapped in fairly strong thick paper. The ends of the paper shall be neatly folded in and affixed by means of gum or other adhesive.

c. The paper cover shall be further secured by means of strong twine or thread both above and across the container, and the twine or thread shall then be fastened on the paper cover by means of sealing wax on which there shall be test four distinct and clear impression of the seal of the sender of which one shall be at the top of the packet, one at the bottom and the other two on the body of the packet. The knots of the twine or thread shall be covered by means of sealing wax bearing the impression of the seal of the sender.

(iv) **Form of order:** The order to be given in writing by the Seed Inspector shall be in Form III

(v) **Form of receipt of records:** When a Seed Inspector seizes any record, register, document or any other material he shall issue a receipt in Form IV to the person concerned.

(vi) **Samples how to sent to the Seed Analyst:** The container of sample for analysis shall be sent to the Seed Analyst by registered post or by hand in a sealed packet enclosed together with a memorandum in Form V in an outer cover addressed to the Seed Analyst.

(vii) **Memorandum and impression of seal to be sent separately:** A copy of the memorandum and a specimen impression of the seal used to seal the packet shall be sent to the Seed Analyst separately by registered post or delivered to him or to any person authorized by him.

(viii) **Addition of preservatives to samples:** Any person taking a sample of seed for the purpose of analysis under the Act may add a preservative as may be specified from time to time to the sample the nature and quantity of the preservative added shall be clearly noted on the label to be affixed to the container.

(ix) **Nature and quantity of the preservative to be noted on the label:** Whenever any preservative is added to a sample the nature and quantity of the preservative added shall be clearly noted on the label to be affixed to the container.

(x) **Analysis of the sample:** On receipt of the packet, it shall be opened either by the Seed Analyst or by an officer authorized in writing in that behalf by the Seed Analyst, who shall record the condition of the seal on the packet. Analysis of the sample shall be carried out at the State Seed Laboratory in accordance with the procedure laid down by the Central Government.

(xi) **Form of notice:** The notice to be given to the person from whom the Seed Inspector intends to take sample shall be in Form VI.

(xii) Form of report: The report of the result of the analysis shall be delivered or sent in Form VII.

(xiii) Fees: The fees payable in respect of the report from the Central Seed Laboratory shall be Rs.10/- per sample of the seed analyzed.

(xiv) Retaining of the sample: The sample of any seed shall be retained under a cool, dry environment to eliminate the loss of viability and in insect proof or rat proof containers. The containers shall be dusted with suitable insecticides and the storage room fumigated to avoid infestation of samples by insects. The samples shall be packed in good quality containers of uniform shape and size before storage.

11. Miscellaneous

This section deals with records and form of memorandum as mentioned below [Part 11].

(i) Records: A person carrying on the seed business shall maintain the following records, namely:

a. DStock record of seed;

b. record of the sale of seeds;

(ii) Form of Memorandum: - The memorandum to be prepared shall be in Form VIII.

N.B.:

1. A white tag shall be used for foundation seed.
2. A purple tag shall be used for registered seed.
3. A blue tag shall be used for certified seed.
4. Certification shall be valid for the period indicated on the tag provided seed is stored in cool dry environment.

Summary

The Seeds Rules, 1968 under Seed Act, 1966 (Act No. 54 of 1966) were formulated and approved for implementation of Seed Act, 1966.These rules consist of 11 Parts and 39 sections. Salient features of seed rules, 1968 related to scope and coverage, central seed committee, central seed laboratory, seed certification agencies, marking and labeling, requirements, certification of seeds, appeals, qualification and duties of seed analysts and seed inspectors, sealing, fastening, dispatch and analysis of samples and miscellaneous matters have been briefly discussed.

Questions

1. **Explain scope and coverage of seed rules, 1968.**

2. **Describe functions of central seed committee.**

3. **What are the functions of central seed laboratory?**

4. **Discuss briefly functions of seed certification agency.**
5. **What are requirements for a person to carry out the seed business ?**
6. **Describe briefly various points related to marking and labeling of seeds.**
7. **Explain various points related to seed certification.**
8. **Describe briefly Qualification and Duties of Seed Inspectors.**
9. **Discuss briefly Qualification and Duties of Seed Analysts.**
10. **Explain the Action to be taken by the Seed Inspector if a complaint is lodged with him.**
11. **Describe briefly the procedure of sealing, fastening, dispatch and analysis of samples.**
12. **Explain the procedure of making appeal against quality of seed.**

Chapter 4

Seed Amendment Rules

Introduction

In India, the Seed Act was first introduced in 1966 which is called as Seeds Act 1966. This Act was amended in 1972, 1973, 1974 and 1981. A brief discussion of these amendments is presented as follows:

1. The Seeds (Amendment) Act, 1972

The first amendment to the seed act 1966 was approved by the central government on 9th September, 1972. This is called the seed amendment Act, 1972. In this act three sections (Section 2, 9 and 25) were amended and one new section (Section 8 A to E) was added. These are explained as follows;

- **(i)** In section 2 (definitions), Jute seeds were inserted after seeds of cattle fodder.
- **(ii)** In section 9 (grant of certificate by certification agency), it was inserted that the seed should meet prescribed standard of germination and purity (instead of minimum limit of germination and purity as given in Seed Act, 1966).
- **(iii)** In section 25 (power to make rules), it should be inserted "the standard to which seed should conform (sub-section 2 after clause f) and the word two successive seasons should be replaced by two or more successive seasons.
- **(iv)** Insertion of new section (Section 8 A to E): This section has provision for establishment of Central seed Certification Board which will advise central government on all matters of seed certification and coordinate the functioning of seed certification agencies. The board will consist of a Chairman (to be nominated by central government), four members from department of agriculture (Not less than director of agriculture), three

members from agriculture universities (not less than director of research), ten members related to seeds and three members as representatives of seed producer. The membership is for a period of two years. The Central Government shall appoint a person as the Secretary of the Board.

Table 4.1: Summary of Seed Amendments

Sl.No.	*Seed Amendment*	*Major amendments*
1.	Seeds (Amendment) Act, 1972.	**(i)** Jute seeds were included in seed Act, 1966. **(ii)** Seed should meet prescribed standards of germination and purity. **(iii)** More than two successive seasons. **(iv)** Provision for establishment of Central seed certification board was added.
2.	Seeds (Amendment) Rules, 1973.	**(i)** Judicial powers of authority have been omitted. **(ii)** Samples to be analyzed as per seed testing manual of ICAR. **(iii)** Testing report to be submitted in not more than 30 days. **(iv)** The word competent authority is replaced by State government.
3.	Seeds (Amendment) Rules, 1974.	If there is crop failure due to poor seed quality, Inspector has to take action against seed supplier on complaint of farmers.
4.	Seeds (Amendment) Rules, 1981.	The Certification agency shall, before granting the certificate, ensure that the seed conforms to the prescribed standards.
5.	Amendments to Seeds Bill, 2004.	Chairpersons of PPV and FR Act and Biodiversity Act were included in the Central Seed Committee.

2. The Seeds (Amendment) Rules, 1973

These amendment rules were approved by the government of India, Ministry of Agriculture (Department of Agriculture) on 30th June, 1973 and may be called Seed Amendment Rules, 1973 of Seed Rules, 1968. The important amendments are as follows.

1. In rule 19 of the Seeds Rule, 1968 (under appeal), the judicial powers of authority have been omitted.
2. In rule 21 of the said rules for sub-rules (2) and (3) the following sub-rules shall be substituted, namely:

   "(2) The Seed Analyst shall analyze the samples in accordance with the procedures laid down in the Seed Testing Manual published by the Indian Council of Agricultural Research as amended from time to time."

   "(3) The Seed Analyst shall deliver in Form VII, a copy of the report of the result of analysis to the persons specified in sub-section (1) of Section 16, as soon as may be but not later than 30 days from the date of receipt of samples sent by the Seed Inspector under sub-section (2) of the Section 15".

4. In rule 23 of the said rules, in clause (h) for the words competent authority "the words" State Government shall be substituted.

3. The Seeds (Amendment) Rules, 1974

These amendment rules were approved by the government of India, Ministry of Agriculture (Department of Agriculture) on 29th April, 1975. These Rules may be called Seed Amendment Rules, 1974 of Seed Rules, 1968. The important amendments are as follows.

After rule 23 of the said rules, the following rule shall be inserted namely "23-A. Action to be taken by the Seed Inspector if a complaint is lodged with him.

(1) If farmer has lodged a complaint in writing that the failure of the crop is due to the defective quality of seeds of any notified kind or variety supplied to him, the Seed Inspector shall take in his possession the marks or labels, the seed containers and a sample of unused seeds to the extent possible from the complaint for establishing the source of supply of seeds and shall investigate the causes of the failure of his crop by sending samples of the lot to the Seed Analyst for detailed analysis at the State Seed Testing Laboratory. He shall thereupon submit the report of his findings as soon as possible to the competent authority.

(2) In case, the Seed Inspector comes to the conclusion that the failure of the crop is due to the quality of seeds supplied to the farmer being less than the minimum standards notified by the Central Government,can launch proceedings against the supplier for contravention of the provisions of the Act or these Rules."

Corrigendum [No. 7-15/74-SD New Delhi, dated the 30th/31st Jan., 1976]

In the Seeds (Amendment) Rules, 1974, published with the notification of the Government of India in the Ministry of Agriculture and Irrigation (Department of Agriculture) (No.GSR 211 (E), dated the 29th April, 1975, in the Gazette of India Extraordinary, Part II, Section 3, Sub-section (i), dated the 29th April, 1975:- at page 863

(i) in line 3 for "to" read "the";

(ii) in line 11, for "makes" read "marks";

(iii) in line 20 and 21, for "the Central Government launch proceedings", read "the Central Government, shall launch proceedings".

4. The Seeds (Amendment) Rules, 1981

These amendment rules were approved by the government of India, Ministry of Agriculture (Department of Agriculture) on 10th June, 1981. These Rules may be called the Seed Amendment Rules, 1981 of Seed Rules, 1968. The important amendments are as follows.

(1) These rules may be called the Seeds (Amendment) Rules, 1981.

(2) They shall come into force on the date of their publication in the Official Gazette. After rule 17 of the Seeds Rules, 1968, the following rule shall be inserted, namely:

17-A The Certification agency shall, before granting the certificate, ensure that the seed conforms to the standards laid down in the Manual known as "Indian Minimum Seed Certification Standards" published by the Central Seed Committee, as amended from time to time."

5. Amendments to Seeds Bill, 2004

In March, the Cabinet had approved the Seeds Bill, 2004 that seeks to regulate the quality of hybrid seeds and check the sale of spurious seeds in the country, besides increasing private participation in seed production and distribution. Amendments to Seed Bill 2004 were approved by the central government on 20th October, 2010.

1. The additional amendments provided for submission of seed related periodic returns to state governments and enhancement in penalties of offences.
2. The government has also included a provision to nominate chairperson of Protection of Plant Varieties and Farmers Rights Authority and National Bio-diversity Authority to the Central Seed Committee.
3. The Bill seeks to repeal and replace existing Seeds Act, 1966, for it does not deal with the quality control of GM seeds, as they are generally not notified.
4. The legislation seeks to regulate the quality of seeds and planting materials, curb the sale of spurious and poor quality seeds, increase private participation in seed production and distribution, and liberalize imports of seeds.
5. Provisions of labeling, seed health, expected performance and compensation to farmers have been included to ensure public accountability.
6. Innovations include compulsory registration, enabling government to exclude certain varieties of seeds on ground of public health, environment, provision for expected performance, seed health and farmer's compensation, etc.

Summary

In India, the Seed Act was first introduced in 1966 which is called as Seeds Act 1966. This Act was amended in 1972, 1973, 1974 and 1981. Amendments of seed act 1972 has provision for inclusion of Jute seeds were included in Seed Act, 1966, seed should meet prescribed standards of germination and purity, and provision for establishment of Central seed certification board.

In the Seed Amendments Rules 1973 Judicial powers of authority have been omitted.

Samples to be analyzed as per seed testing manual of ICAR, seed testing report is to be submitted in not more than 30 days and the word competent authority is replaced by State government.

As per seed Amendments rules 1974, If there is crop failure due to poor seed quality, Seed Inspector has to take action against seed supplier on complaint of farmers.

As per seed Amendments rules 1981, the Certification agency shall, before granting the certificate, ensure that the seed conforms to the prescribed standards.

Amendments to Seeds Bill, 2004, includes Chairpersons of PPV and FR Act and Biodiversity Act in the Central Seed Committee.

Questions

1. **Describe briefly Seed amendment act 1972.**
2. **Explain briefly Seed amendment rules 1973.**
3. **Discuss briefly Seed amendment rules 1974.**
4. **Discuss briefly Seed amendment rules 1981.**
5. **What are major changes suggested by Seed amendment act 1972.**
6. **Explain major amendments of Seed amendment rules 1973.**
7. **Discuss briefly main changes suggested by Seed amendment rules 1974.**
8. **Discuss briefly major amendments to Seed Bill 2004.**
9. **Write short notes on the following:**
   (i) Seed amendment act 1972 **(ii)** Seed amendment rules 1973.

Chapter 5

Seeds (Control) Order, 1983

Introduction

The Seeds (Control) order, 1983 under Seed Act, 1966 (Act No. 54 of 1966) was approved by the government of India, Ministry of Agriculture (Department of Agriculture and Cooperation) on 30th December, 1983 and may be called Seed control Order, 1983. This order consists of four parts. Part 1 deals with title, scope and definitions of various terms. Part 2 deals with licensing of dealers, part 3 deals with appointment of licensing authority, appointment of inspectors, inspection and punishment, time limit for analysis, suspension/cancellation of license and appeals. Part 4 deals with amendment of license, maintenance of records and submission of returns, etc. These are briefly presented as follows.

1. Title and Scope

This order may be called the Seeds (Control) Order, 1983. It extends to the whole of India. It shall come into force on the 30th December of 1983. The central Government hereby declare the following seeds used for sowing or planting (including seedlings and tuber, bulbs, rhizomes, roots, cuttings and all types of graft and other vegetatively propagated material of food crops or cattle fodder) to be essential commodities for the purpose of the said Act, namely.

(i) Seeds of food crops and seeds of fruit and vegetables.

(ii) Seeds of cattle fodder and

(iii) Jute seeds.

2. Compulsory Licensing

According to this seed order, it is compulsory for seed dealers to obtain license to carryout seed business. No person shall carry on the business of selling, exporting or importing seeds at any place except under and in accordance with the terms and conditions of license granted to him under this Order. Various aspects of licensing are briefly presented as follows.

Sl.No.	*Particulars*	*Brief Explanation*
1	Scope and Coverage	This seeds order extents to whole of India and covers seeds of food crops, fruits and vegetable crops, fodder crops and jute.
2	Compulsory Licensing	This section deals with various aspects of licensing such as grant or refusal of license, renewal of license, period of validity of license, etc.
3	Licensing Authority	This section deals with appointment of licensing authority, seed inspectors, powers of seed inspectors, time limit for analysis report, cancellation or suspension of license, appeal etc.
4	Amendment of License	This section deals with amendment of license, maintenance of records and submission of monthly returns.

(i) **Application for License:** Every person desiring to obtain a license for selling, exporting or importing of seeds shall make an application in duplicate in Form "A" together with a fee of Rs.50/- for license to licensing authority.

(ii) **Grant and Refusal of License:** The licensing authority may, after making such enquiry as it thinks fit, grant a license in Form "B" to any person who applies for it. However, issue of license to a person can be refused under following conditions.

a. If his earlier license granted under this Order is under suspension. The refusal would for the period of such suspension.

b. If his earlier license granted under this Order has been cancelled. The refusal would for a period of one year from the date of cancellation.

c. If a person has been convicted under the Essential Commodities Act, 1955(10 of 1955) or any order issued there under within three years preceding the date of application.

When the licensing authority refused to grant license to a person who applied for it, he shall record his reasons for doing so.

(iii) **Period of validity of license:** Every license under this order, shall, unless previously suspended or cancelled, remain valid for three years from the date of its issue.

(iv) **Renewal of License:** For renewal of license, the license holder has to give an application in the prescribed form (Form C) to the licensing authority before expiry of the license. The fee of rupees twenty is to be paid for renewal. If any application for renewal is not made before the expiry of the license, but is made within one month from the date of expiry of the license, the license may be renewed on payment of additional fee of rupees twenty-five, in addition to the fee for renewal of license.

(v) **Display of Stock and Price List:** Every seed dealer has to daily display the opening and closing stocks, of different seeds held by him and a price list of different seeds available with him.

(vi) **Issue of Cash/Credit Memo to Purchaser:** Every dealer shall give a cash or credit memorandum to a purchaser of seeds.

3. Licensing Authority, etc.

For effective implementation of the order, the State Government has to appoint licensing authority and inspectors areas wise and delegate them certain powers. The power of inspection and punishment are also decided. A seed inspector has the following powers.

(i) He can ask any dealer to give information about purchase, storage and sale of seeds by him;

(ii) He can enter and search any premises where any seed is stored or exhibited for sale to ensure compliance with the provisions of this Order;

(iii) He can draw samples of seeds meant for sale, export and seeds imported, and send the same to a laboratory to ensure that the sample conforms to standard of quality claimed;

(iv) He can seize or detain any seed in respect of which he has reason to believe that contravention of this order has been committed or is being committed;

(v) He can seize any books of accounts or document relating to any seed in respect of which he has reason to believe that a contravention of this Order has been committed and is being committed. Every seed dealer will have to cooperate with the seed inspector in this matter.

Duties of Inspector

The Inspector shall give receipt, in respect of the books of accounts or documents seized, to the person from whom they have been seized.

The seized books of account or documents shall be returned to the person from whom the same had been sized after copies thereof or extracts there from as certified by such person have be taken.

When any seed is seized by an Inspector, he shall forthwith report the fact of such seizure to a Magistrate in his jurisdiction for further action.

Time Limit for Analysis

The laboratory to which a sample has been sent by an Inspector for analysis under this order shall analyze the said samples and send the analysis report to the concerned Inspector within 60 days from the date of receipt of the sample in the laboratory.

Suspension/Cancellation of License

If the license has been obtained by misrepresentation or any provision of the order/condition of the license has been contravened, the licensing authority may suspend or cancel the license. However, the license holder is given an opportunity to clarify before such action.

Appeal

If any person has been refused to grant, amend or renew the license for sale, export or import of seeds or his license has been cancelled/suspended, he may appeal within sixty days from the date of the order, to such authority as the State Government may specify in this behalf, and the decision of such authority shall be final. The appeal application shall accompany an appeal fee of rupees fifty.

4. Amendment of License

The licensing authority may, on receipt of a request in writing together with a fee of rupees ten from a dealer, amend the license of such dealer.

Maintenance of Records and Submission of Returns etc.

Every dealer shall maintain such books, accounts and record relating to his business as may be directed by the State Government. Moreover, every dealer shall submit monthly return relating to his business for the preceding month in Form "D" to the licensing authority by the 5th day of every month.

Summary

The Seeds (Control) order, 1983 under Seed Act, 1966 (Act No. 54 of 1966) was approved by the government of India, Ministry of Agriculture (Department of Agriculture and Cooperation) on 30th December, 1983 and may be called Seed control Order, 1983. This order consists of four parts. Part 1 deals with title, scope and definitions of various terms. Part 2 deals with licensing of dealers, part 3 deals with appointment of licensing authority, appointment of inspectors, inspection and punishment, time limit for analysis, suspension/cancellation of license and appeals. Part 4 deals with amendment of license, maintenance of records and submission of returns, etc. These have been amply discussed.

Questions

1. **Explain title and scope of seed control order, 1983.**
2. **Describe various aspects of licensing for seed dealers.**
3. **Discuss validity and renewal of license for seed dealers.**
4. **Write short notes on the following:**
   - **(i)** Appointment of licensing authority
   - **(ii)** Appointment of seed inspectors
   - **(iii)** Inspection and punishment
   - **(iv)** Time limit for seed analysis

5. Explain the following:

(i) Suspension/cancellation of license

(ii) Amendment of license

6. Define the following terms:

(i) Seed export

(ii) Seed import

(iii) Dealer

(iv) Seed inspector

(v) Registering authority

Chapter 6

New Seed Policy, 1988

Introduction

The New Seed Policy was approved by the Government of India, Ministry of Agriculture [Department of agriculture and Cooperation] on 16th September, 1988, This seed policy has been approved with the following objectives.

1. To provide the best planting materials to the farmers available in the world.
2. To increase agriculture productivity and thereby increasing farm income.
3. To increase export earnings through planting material.

Materials Covered

The Policy covers the import of following plant materials:

1. Seeds of wheat and paddy;
2. Seeds of coarse cereals/pulses/oil seeds;
3. Seeds of vegetable and flower;
4. Bulbs/tubers of flowers;
5. Cuttings/saplings/bud/wood etc., of flowers; and
6. Seeds and planting material of fruits.

Guidelines

The following are the guidelines for the import of above categories of seed/planting material under the Seed Policy.

1. Import of Seeds of Coarse Cereals/Pulses/Oilseeds

It is necessary to increase the production of coarse cereals, oilseeds and pulses by making available high yielding seeds to the farmers. At the same time, repetitive bulk imports of these seeds cannot be permitted indefinitely. Based on these considerations, the new import policy for seeds for sowing of coarse cereals, oilseeds and pulses is detailed in the following paragraphs. Agreement to supply parent line seeds/breeder seeds/technology NSC, SSC's may import Issue of Import License by the CCI and E Seed for trial evaluation by ICAR Recommendation of DAC Publicity through Media.

(i) **Period of Import:** The import of seeds of coarse cereals, pulses and oilseeds for sowing may be allowed for a period not exceeding two years by companies, which have technical/financial collaboration agreements for production of seeds with companies abroad, provided the foreign supplier agrees to supply parent line seeds/nucleus or breeder seeds/technology to the Indian company within a period of two years from the date of import of the first commercial consignment after its import has been recommended by DAC.

(ii) **Organization to Import:** The National Seeds Corporation and the State Seeds Corporation may be permitted to import of coarse cereals, oilseeds and pulses on the recommendations of DAC.

(iii) **Grant of Import License:** Plant Protection Advisor (PPA) or any other officer notified for the said purpose issues the import permit based on the recommendation of DAC. DAC may, within 30 days of the receipt of the application, reject the application or recommend it to the Central Committee for Import and Export (CCI and E) for grant of an import license to the importer. The CCI and E shall issue the import license within 15 days of the receipt of the recommendations from DAC. However the recommendation of DAC is not required for import of trial material.

(iv) **Quantity of Import:** A quantity of 5 kg of seed of coarse cereals/pulses and oilseeds for sowing sought to be imported by eligible importers, will be given to ICAR or such farms, which are accredited by ICAR, for trial and evaluation. After trial/evaluation for one crop season, the ICAR will intimate the results, agro-climatic zone-wise, within three months of the season to DAC. After the receipt of the results of the ICAR trial/evaluation, an eligible importer may apply for the import of such seed to the DAC.

On receipt of the application for commercial import, the DAC will consider the trial/evaluation report on the performance/resistance to seed and soil borne diseases of the seeds proposed to be imported.

(v) The characteristics of the varieties/hybrids cleared by DAC for import shall be 'given publicity through the AIR, Doordarshan and the extension agencies so that farmers will get to know the potential of the varieties/hybrids cleared for import, as 'compared to the indigenous varieties/hybrids. The decision to import or not to import the varieties/hybrids cleared by DAC will be left to the importers.

(vi) **Seed for Gene Bank:** All importers shall make available a small but specified quantity of the seed being imported to the ICAR at cost price, for testing/ accession to the Gene Bank with the NBPGR.

(vii) The bulk import of seeds of coarse cereals, oilseeds and pulses shall be cleared/rejected by the PPA after quarantine checks within a period not exceeding three weeks. The rejected consignment shall be destroyed in accordance with the prescribed procedure.

(viii) The imported consignment shall be subjected to detailed testing for a period of 30-35 days. During quarantine the imported consignment shall be kept in a customs bonded warehouse at the cost of the importer.

2. Import of Vegetable and Flower Seeds

The availability of high yielding vegetable seeds is essential for increasing vegetable production and productivity. The additional production will generate higher incomes to the farmers and ensure the increased availability of vegetables at reasonable prices to the consumers.

(i) Further, there is a significant demand for fresh vegetables and flowers in the international market. The production, on a competitive basis, of vegetables and flowers of international standards has to be developed/ accelerated to increase our export earnings.

(ii) It is, therefore, necessary to supply to the farmers the best available seeds of vegetables/flowers and ornamental plants. Accordingly, the import of seeds of vegetables/flowers and ornamental plants, tubers and bulbs, cuttings, saplings, bud woods, etc. of flowers will be allowed on OGL by identified categories.

(iii) The bulk import of vegetable and flower seeds is permitted under Open General license (OGL). The following organizations or importers shall be eligible to import these seeds/planting materials.

  (1) Departments of Agriculture/Horticulture of the State Government, State Agriculture Universities and ICAR;

  (2) Seed producing Indian companies/firms, after registration with the National Seeds Corporation;

  (3) National Seeds Corporation, State Seeds Corporations;

  (4) Food processing industrial units;

  (5) Growers of vegetables and flowers registered with the Director of Agriculture/Horticulture of the State Government.

(iv) No post entry quarantine (PEQ) shall be necessary for the import of seeds of vegetables/flowers/ornamental plants and tubers and bulbs of flowers. The import will be permitted after visual inspection, fumigation, laboratory inspection and grow-out tests.

(v) The Plant Protection Adviser to the Government of India (PPA) may reject/ clear the consignment of seeds of vegetables/flowers/ornamental plants and tubers and bulbs of flowers within a period not exceeding 50 days. The consignment shall be kept in a bonded warehouse at the cost of the importer until quarantine/customs clearance is given. The customs clearance of seeds of vegetables/flowers/ornamental plants and tubers and bulbs of flowers will be given by the customs authorities only after the quarantine clearance is given by the PPA.

The imported seed is subject to detailed seed testing for a period of 30-35 days on arrival at the port of entry. However, permit issued by PPA or any other competent officer notified for this purpose, is required for import of vegetable seeds but not flower seeds.

3. Import of Bulbs and Tubers of Flowers

(i) The bulk import of bulbs and tubers of flowers and ornamentals is allowed under OGL by eligible importers as stated above.

(ii) The imported consignments shall be subjected to grow-out test for a period of 35-40 days on arrival at the port of entry and for the same purpose the consignment will be held under detention in cold storage under customs bond or the imported bulbs and tubers may be subjected to post-entry quarantine instead of grow-out test at the specific request of the importer.

(iii) The bulbs/tubers are required to be sown in individual poly bags and are subjected to joint inspection by designated Inspection Authority (DIA) and the officer of concerned plant quarantine (PQ) station during post entry quarantine (PEQ) period specified at the time of issue of permit.

4. Import of Cuttings/Saplings/Bud Wood etc., of Flowers

(i) The import of cuttings/saplings/bud wood etc., of flowers is permitted for import under Open General license (OGL).

(ii) A permit issued by PPA or any other competent officer notified for this purpose, is required. The importer is required to establish PEQ facilities prior to import, which are to be approved by the Designated Inspection Authorities (DIAs) as per the guidelines issued by PPA.

(iii) The imported consignments on arrival at the port of entry shall be subject to quarantine inspection and cleared within 24-72 hrs with a condition for growing under post-entry quarantine for a period not exceeding 45 days in an approved PEQ facility under the supervision of DIA.

For import of cuttings, saplings, bud wood etc. of flowers, PEQ shall be essential. PEQ facilities as may be prescribed by the PPA, shall have to be established by the eligible categories of importers permitted to import cuttings, saplings, bud wood etc. of flowers under OGL. The fact that such a facility has been established by him under the Plants, Fruits and Seeds (Regulation of import into India) Order, 1984. This agency, hereinafter referred to as the designated inspection agency (DIA), which may be the State Agriculture University (SAU), will also be authorized to carry out the PEQ inspection. The DIA's certificate that the importer has established the prescribed PEQ facility shall be presented by the eligible importer to the PPA before the imported consignment is given quarantine clearance by the PPA.

(i) The quarantine clearance subject to the PEQ condition will be given by the PPA after conducting visual inspection/laboratory inspection to see that the consignment of the imported cuttings, saplings, bud wood etc. of flowers, is free from any exotic pests or diseases.

(ii) Customs clearance shall be given by the customs authorities to the consignment on receipt of the quarantine clearance with PEQ condition from PPA. bulbs of flowers.

Approval of PEQ facility Guidelines to designated inspection agency for PEQ Samples to ICAR/Gene Bank Modification in REP Policy Import Policy for Seeds/Planting Introduction of pests and pathogens.

(iii) Cuttings, saplings, bud woods etc. of flowers, which require PEQ inspection shall be grown in the PEQ facility approved by the PPA or a DIA authorized by the PPA. The period for which the cuttings, saplings, bud woods etc. shall be grown in such a facility shall be specified in the import permit to be granted by the PPA.

(iv) Detailed guidelines for the DIA (Appendix A) for conducting post entry quarantine shall be laid down by the PPA from time to time. If the PEQ reveal any exotic insect pest or disease the material shall be forthwith destroyed in the prescribed manner by the DIA under intimation to PPA.

(v) Small quantities of n flower/vegetable seeds tubers and cuttings and saplings will be made available by the importers to ICAR at cost price for testing/accession to the Gene Bank.

(vi) The import of seeds of vegetables, flowers, ornamental plants and bulbs and tubers of flowers, will continue to be allowed under REP. However, the transfer of REP licenses for items of seeds of vegetables/flowers/ornamental plants, bulbs and tubers of flowers will be allowed only to those persons, who are otherwise eligible to import these items under OGL subject to the conditions laid down in the Plants, Fruits and Seeds (Regulation of Import into India) Order, 1984.

(vii) The import of seeds/planting material for fruits shall be permitted selectively by DAC, on a case-to-case basis, on the recommendation of the State Director of Horticulture/Agriculture, subject to such quarantine regulations as may be laid down by the PPA.

5. Import of Seeds/Planting Material of Fruits

(i) The seeds/planting material of fruit plant species are selectively permitted for import by DAC on case-to-case basis on the recommendation of Director of Horticulture/Agriculture of the state and subject to quarantine regulations as may be laid down by PPA.

(ii) However, permit issued by PPA or any other competent officer notified for this purpose, is required for the import of the same.

Strengthening of Plant Quarantine

1. In the past, several pests and pathogens have been introduced through seed material imported into the country. Many of these pests have since become wide-spread. Some examples are downy mildew on sunflower, spotted wilt virus on tomato, peanut stripe virus on groundnut, apple scab from Europe, golden nematode and wart diseases of potato from Europe and bacterial blight of paddy from East Asia, to mention only a few.
2. Therefore, while importing seeds and planting materials, it has to be ensured that there is absolutely no compromise on plant quarantine procedures. Every effort has to be made to prevent the entry into India of exotic pests, diseases and weeds, which are detrimental to the interests of the farmers. The new policy on imports for seeds can thus be implemented fully only after the plant quarantine facilities are adequately strengthened within a specified time frame.
3. There are 26 quarantine and fumigation stations located at 10 airports, 9 seaports and 7 land frontiers. It is proposed to allow import of seed/planting material etc. initially only at Delhi, Bombay, Calcutta, Madras and Amritsar.

   Therefore, the plant quarantine facilities at these five centres are being strengthened/modernized.
4. The strengthening of the plant quarantine organization through the augmentation of technical staff, provision of physical infrastructure, essential laboratory equipment and the creation of regional offices for issue of import permits shall be a time-bound operation and shall be completed within a period of 12 months.
5. The new policy on import of seed shall come into effect from the 1st October, 1988.
6. The imports under the new policy shall be allowed, based on the procedures being followed by PPA and subject to the existing physical facilities of the Plant Quarantine and Fumigation Stations to undertake laboratory tests, fumigation etc. for a period of three months *i.e.* till the 31st December, 1988.
7. During this 3 months period, PPA, while strengthening the quarantine facilities for grow-out tests available at the IARI, New Delhi and the SAUs or other research laboratories near Bombay, Calcutta, Madras and Amritsar. By the 1st of October, 1999, based on the time-bound program, the plant quarantine facilities at the 5 stations should have been augmented to implement the new policy in full. Pending this, from the 1st January to the 1st October, 1989, the implementation of the new policy will be phased/regulated, based on the progress in the commissioning of the new plant quarantine facilities at these five stations.
8. A National Plant Quarantine Advisory Committee shall be constituted to advise the PPA on various technical matters in different disciplines of Entomology, Plant Pathology, Nematology, Virology, Bacteriology, Weed Sciences and for serological detection techniques of viruses. This Committee may meet at least twice a year.

Guidelines and Procedures for Quarantine/Post Entry Quarantine

1. A separate Plant Quarantine Counter shall be provided in the arrival hall at Delhi, Bombay, Calcutta, Madras and Amritsar airports and at Bombay, Calcutta and Madras sea ports for effectively checking the imported seeds and planting materials and ensuring strict observance of plant quarantine procedure.
2. All consignments of seeds and plants for sowing shall be imported into India in accordance with the provisions of Plants, Fruits and Seeds (Regulation of Import into India) Order, 1984. An import permit and a phytosanitary certificate with additional declaration, if necessary, as prescribed by the PPA, shall be essential. The pre-entry requirements for flowers and ornamental plants, vegetable seeds (FVS), bulbs and tubers (BT) and cuttings and saplings (CS) would be as follows:
   (i) Import permit
   (ii) Phytosanitary Certificate
   (iii) Additional Declaration
      (a) Freedom from Soil
      (b) Freedom from Weeds
      (c) Plant Quarantine objects and
   (iv) Approval of PEQ facilities
3. After the arrival of consignments at the port of entry, quarantine checks would be undertaken which may include visual inspection, laboratory inspection, fumigation and grow out tests. For the purpose of these checks samples will be Post Entry Quarantine. Sampling Procedures: Inspection requirements drawn and the tests will be conducted concurrently.
4. PEQ will be required only for growing imported cuttings, saplings, budwood etc. of flowers. The importers will be required to create PEQ facilities, as may be prescribed by the PPA. The guidelines for PEQ facilities for importers are at Appendix 'B'. The planting materials, which require PEQ inspection, shall be grown in the PEQ facility approved by the PPA or the DIA in the State, authorized by the PPA under the provisions of the Plants, Fruits and Seeds (Regulation of Import into India) Order, 1984. The period, for which the planting materials shall be grown in such a facility, shall be specified in the import permits granted by PPA. On the production of the certificate from the DIA that the importer has established the prescribed PEQ facility, the PPA will give the quarantine clearance after visual tests, laboratory tests, etc.
5. The sampling procedures for various categories of imported vegetable/flower planting materials would be as under:
   (a) FVS The sampling techniques shall be as per International Seed Testing Association (ISTA) rules and guidelines.

(b) BT and CS The minimum sample size is considered to be 10 numbers. However, in case of bulk imports, the representative sample for V.I. and L.I. would not be less than 0.1 per cent.

6. The inspection under the categories mentioned above would comprise the following specific investigations/parameters.
7. The technique for undertaking grow out tests has been prescribed by the PPA. The seeds of vegetables/flowers/ornamental plants after visual inspection shall be subject to fumigation at the cost of the importer to check pest infestation/disease before undertaking laboratory inspection.
8. If the imported consignment shows the presence or manifestation of any exotic disease, pests or nematodes, the entire consignment shall be rejected and disposed of in accordance with the prescribed methods of destruction.
9. The imported consignment showing presence of pests, diseases and nematodes recorded in India shall be released only after effective quarantine measures are undertaken to bring down the incidence to levels prescribed.

   otherwise, such consignments shall be disposed of as prescribed by the PPA. 6.10 If the samples drawn from the imported consignments satisfy the prescribed checks, the consignments shall be released for entry.

Incentives to Domestic Seed Industry

1. In the context of the liberalization of the import of seed/planting materials, incentives are proposed to be provided to domestic seed industry to encourage and promote its growth and enable it to produce viable high yielding varieties.
2. Accordingly, it has been decided to reduce the import duty on seeds to 15 per cent. The import duty reduction concession shall be accorded only to seeds meant for sowing and to planting materials.
3. It is proposed to reduce the import duty on machines and equipments used for seed production/processing, quality control machinery etc., not manufactured in the country or for which the technology upgrading is necessary.
4. Pre-shipment credit up to 180 days will be allowed at 9.5 per cent rate of interest per annum. Beyond 180 days, and in all up to 270 days with the prior approval of the Reserve Bank of India, the rate of interest shall be 11.5 per cent per annum.

   Post-shipment credit shall also be allowed at 9.5 per cent rate of interest per annum. A circular specifically allowing these facilities to seed exploring units will be issued by the Reserve Bank of India.
5. Cash compensatory support on the export of seed will be considered by the Ministry of Commerce on the basis of data supplied by the seed exporting companies.

6. Income tax rebate/deduction will be given to the seed industry on the revenue expenditure incurred on in-house research and development, to such units which pay income tax.
7. Incentives, as provided to industries located in the backward areas/growth centres, may be, sought by DAC for seed industry. Such concessions for seed industry will be identified by DAC and taken up with the Ministry of Finance.
8. A High Level Review Committee in the DAC will monitor the progress of the implementation of the Policy.
9. The New Policy on Seed Development should help the seed industry to grow. The farmers will be benefited by getting access to the best available seed and planting material available anywhere in the world.
10. The Government of India trusts that the State Governments./ Administrations of Union Territories/State Agricultural Universities will extend their fullest cooperation and assistance in the implementation of the Policy.

APPENDIX–A: Guidelines for Designated Inspection Agencies under the Plants, Fruits and Seeds (Regulation of import into India) Order, 1984

1. Post Entry Quarantine (PEQ) shall be essential for imported cuttings, saplings, bud woods, etc. of flowers.
2. Eligible categories of importers permitted to import cuttings, saplings, bud woods, etc. of flowers must establish post entry quarantine facilities, as may be prescribed by the PPA.
3. A certificate shall be issued by the PPA or the Designated Inspection Agencies in the States authorized by the PPA under the Plants, Fruits and Seeds (Regulation of Import into India) Order, 1984 for conducting PEQ inspections, that the importer has established the prescribed PEQ facilities, after verification and evaluation of the PEQ facilities created by the importer.
4. Imported cuttings, saplings, bud woods etc. of flowers shall be grown in the prescribed PEQ facility for a period specified in the import permit granted by the PPA.
5. The DIA shall be located in the State Agricultural University.
6. A multidisciplinary team of the DIA consisting of a Plant Pathologist/ Virologist/Entomologist shall inspect the imported planting material at the time of planting in the specified PEQ facility.
7. Post planting observations shall be taken in the following manner: Cuttings: The first observation would depend upon the growth of the plant but normally it should be undertaken within 20-25 days followed by second observation within 40-45 days of planting.

   Saplings: The first observation after planting shall depend upon the growth of the plant, which may normally be within 2-3 weeks.

During inspection, special notice should be taken of symptoms of any plant diseases, especially systemic and viral disease.

Disease affected plants should be destroyed in the presence of the DIA, as prescribed by PPA, to avoid any possible contamination by pathogens of the surrounding environment and plants.

In the event of observance of any exotic pest or disease, the material shall be forthwith destroyed in the prescribed manner under intimation to PPA.

8. The final report of PEQ inspection shall be sent to PPA for information and record within 15 days of the final observation.

APPENDIX-B: Guidelines for Importers for Establishing Post Entry Quarantine Facilities

1. The importers of cutting and saplings, etc. of flowers will be required to establish post entry quarantine facilities, as prescribed by the Plant Protection Adviser. These facilities may include the establishment of a glass house/plastic house/any other specified in a defined area, in isolation.
2. Imported cuttings, saplings, bud woods, etc. of flowers shall be grown in the Prescribed PEQ facility for a period specified in the import permit granted by the PPA.
3. The importer shall facilitate the Designated Inspection Agency (DIA) to undertake inspection at specified intervals of time.
4. The importer shall inform the DIA about the planting dates well in advance. If there is any occurrence of pest or disease symptoms, he shall immediately inform the DIA. The importer shall abide by the decision of the DIA.
5. No plant or parts thereof whether living or dead shall be removed during the pendency of post entry quarantine without the approval of the DIA.
6. A qualified Plant Pathologist/Entomologist/Plant Virologist/ Horticulturist should be engaged as consultant at the cost of the importer for looking after the plants under post entry quarantine.
7. On the detection of any exotic insect pest or disease, the entire imported material shall be destroyed forthwith in a prescribed manner under intimation to PPA.
8. During inspection, the multidisciplinary team of the DIA shall take special notice of symptoms of any plant diseases specially systemic and viral diseases. The disease affected plants shall be destroyed in the presence of DIA to avoid possible contamination of the surrounding plants and environment by pathogens and the soil shall be disinfected with formaline.
9. The importer shall pay the cost of post entry quarantine inspection, as prescribed by PPA. He shall make the deposit, as may be laid down, on the quarantine clearance of the consignment by the PPA. The consignment shall be released only after the importer makes the payment in full of the DIA inspection fees to the PPA.

10. The importer shall maintain basic inspection tools like hand lenses, microscopes, chemicals, dissection box, glass vials and petri plates.
11. A log book shall be maintained for proper record for incoming and out going material as well as the dates of inspection and detailed report by the DIA.
12. A register containing details of imported material, panting date, manifestation of diseases and appearance of pests and control measures taken shall be maintained.
13. The PEQ areas shall have the provision of locking arrangements for security purposes.

Summary

The New Seed Policy was approved by the Government of India, Ministry of Agriculture [Department of Agriculture and Cooperation] on 16th September, 1988, This seed policy has been approved with three main objectives, *viz.* (i) to provide the best planting materials to the farmers available in the world, (ii) to increase agriculture productivity and thereby increasing farm income and (iii) to increase export earnings through planting material.

The New Seed Policy, 1988 covers the import of seeds of wheat and paddy, seeds of coarse cereals/pulses/oil seeds, seeds of vegetable and flower, bulbs/tubers of flowers, cuttings/saplings/bud/wood etc., of flowers; and seeds and planting material of fruits. Details about the import of planting materials of all these crops have been presented. Besides these, strengthening of plant quarantine guidelines for designated inspection agencies under the plants, fruits and seeds (regulation of import into India) order, 1984 and guidelines for importers for establishing post entry quarantine facilities have been discussed.

Questions

1. **Describe objectives and materials covered by the New Seed Policy, 1988.**
2. **Explain briefly various points related to import of seeds of coarse cereals/ pulses/oil seeds.**
3. **Explain briefly various points related to import of seeds of coarse cereals/ pulses/oil seeds.**
4. **Discuss in short various points related to import of vegetable and flower seeds.**
5. **Discuss in short various points related of Import of cuttings/saplings/bud wood etc., of flowers.**
6. **Describe briefly various points related to Import of bulbs and tubers of flowers.**
7. **Describe briefly various points related to Import of seeds/planting material of fruits.**

8. **Explain briefly various points related to strengthening of Plant Quarantine facilities.**

9. **Discuss in brief various incentives to promote Domestic Seed Industry in India.**

10. **Describe briefly guidelines for designated inspection agencies under the plants, fruits and seeds (regulation of import into India) order, 1984.**

11. **Give a brief account of the guidelines for importers for establishing post entry quarantine facilities.**

Chapter 7

Seed Order, 1989

Introduction

The seed order, 1989 was approved by the government of India, Ministry of Agriculture (Department of Agriculture and Cooperation) on 27th October, 1989. This order suppresses the Plants, Fruits and Seeds (Regulation of import into India) Order 1984. This order may be called the Plants, Fruits and Seeds (Regulation of Import into India) Order, 1989. This order consists of six chapters. Chapter 1 deals with title, scope and definitions of various terms. Chapter 2 deals with general conditions of import. Chapter 3 explains special conditions of import. Chapter 4 deals with post entry quarantine, Chapter 5 deals with appeals and revision and chapter 6 explains powers of relaxation. These are briefly presented as follows.

1. Title and Scope

This order may be called the Plants, Fruits and Seeds (Regulation of Import into India) Order, 1989. It extends to the whole of India. [Chapter 1].

2. General Conditions for Import

All consignments of plants, fruits and seeds shall be imported into India subject to the following conditions.

(1) **Valid Permit:** A valid permit is compulsory for import of any consignment into India.

(2) **Application:** An application for a permit to import consignments by land, air or sea has to be sent in triplicate) at least one month in advance to the Competent Authority. The application for the import of seeds, fruits and Plants for consumption shall be made in form 'A' and that for the import of seeds and plants for sowing or planting shall be made in form 'B'.

(3) **Fees:** A fee of Rs.50 is to be paid along with the application for the import of seeds, fruits and plants for consumption and Rs.100 for application for the import of seeds and plants for sowing or planting. The fee is to be paid in the form of Demand Draft payable to the Competent Authority having jurisdiction.

(4) **Issue of Permit**: If the application is in order, the Competent Authority shall issue permit in Form "C" for import of seeds and fruits for consumption and in Form "D" for import of seeds and plants for sowing or planting. If application does not meet requirements, the issue of permit may be refused or withheld by the Competent Authority.

(5) **Validity of Permit:** The import permit issued for such import shall be valid for a period of six months. However, the Competent Authority may, on request, extent the period of validity for a further period of six months, for genuine reasons.

(6) **Tag Color:** The Competent Authority shall forward to the importer an orange and green color tag specified in form "E", in the case of permits issued for import of seeds and plants for sowing or planting so as to facilitate the identification of consignments at the time of their arrival at the land customs station or port of entry.

(7) **Entry Points:** All the consignments for consumption, sowing and propagation or planting shall be imported into India only through entry points notified by the Central Government from time to time in this behalf.However, all consignments of dry fruits, fresh fruits and vegetables for consumption, imported from Afghanistan, Pakistan and a West Asian Countries by land shall be imported only through Attari-Wagha Border.

(8) All consignments of plants and seeds for sowing and propagation or planting shall be imported into India through land customs stations, seaport, airport at Amritsar, Bombay, Calcutta, Delhi and Madras and such other entry points as may be specifically notified by the Central Government from time to time.

(9) **Inspection of Material:** The consignment, on arrival, at an entry point, shall be inspected by the Plant Protection Adviser or any other officer duly authorized by him in this behalf, in accordance with the guidelines issued by the Plant Protection Adviser from time to time.

(10) The Plant Protection Adviser or the officer authorized by him may, after inspection, fumigation, disinfection or imported for personal consumption may be allowed to be imported without a Phytosanitary to the country of origin.

(11) **Cost of Fumigation, etc:** Where fumigation or disinfestation or the; disinfection is considered necessary in respect of a consignment of plants, seeds and fruits of more than 1000 cubic metre in volume, the importer shall on his own or at his cost through an agency approved by the Plant Protection Adviser arrange for the fumigation, disinfection or disinfection

of the consignment, under the supervision of an officer duly authorised by the Plant Protection Adviser in that behalf.

(12) Responsibility of Importer: The importer has the following responsibility.

(a) To bring the consignments to the concerned Plant Quarantine and Fumigation Station, or to places of inspection, fumigation or treatment as directed by the Plant Protection Adviser or the officer duly authorized by him;

(b) To open, repack and load into or unload from the fumigation chamber and seal the consignments; and seal the consignments; and

(c) To remove them after inspection and treatment, according to the directions issued by the Plant Protection Adviser or an officer duly authorized by him.

The consignments intended for other countries shall be allowed transit through or transshipment at air or sea ports or land customs stations, provided they are packed in such a manner as will not permit spillage of any soil or materials or escape of any pest, and subject also to the condition that they are not opened in any place in India.

(13) Phytosanitary Certificate: No consignment shall be imported unless accompanied by an official Phytosanitary Certificate issued by the authorized officer of the country of origin of the consignment. Provided that cut flowers, garlands, bouquets, fruits and vegetables weighing less than two kilograms.

(14) Packaging: Consignments for import should be packed in the packaging materials envisaged as in clause 2(g) of this order. No consignment wherein hay or straw or any material of plant origin is used for packaging or as a part of packaging material shall be allowed to be imported.

(15) Import of Soil: Import of soil, earth, compost, sand, plant debris along with plants, fruits or seeds shall not be permitted except under the following conditions:

(a) The consignments of soil, earth, clay and similar material for any microbiological, soil-mechanics or minerological investigation and peat for horticultural may be permitted through specified air or sea ports or land custom station, on applications made for that purpose.

(b) The application for the purpose referred to in (I) above shall be made to the Plant Protection Adviser, at least one month in advance, in form "F"

(c) The Plant Protection Adviser may, after scrutiny of the application, and if satisfied of the purpose for which such consignment is being imported, issue special permit in Form "G".

(d) The consignments shall be inspected, fumigated disinfected or disinfested, on arrival, by the Plant Protection Adviser or any other officer duly authorized by him in this behalf.

(15) Fees: The importer of the consignments or his agent shall pay to the Plant Protection Adviser or any other officer duly authorized by him in this behalf, the fees prescribed in Schedule III to meet the cost of inspection, fumigationand disinfection before the release of the consignments.

3. Special Conditions for Import

In addition to the general conditions of import of Plants, Fruits and Seeds, there are some special conditions for import of these commodities. The following items can be imported [Chapter 3].

(i) Coconut, seeds and all species of Cocos;
(ii) Coffee plants and seeds, and all species of Coffee;
(iii) Cotton seeds, and all species of Gossypium;
(iv) Seeds of forest trees;
(v) Groundnut seeds, and all species of Araches.
(vi) Bucrene and all species of Medicago;
(vii) Potato and all species of Solanum;
(viii) Rubber and all species of Hevea;
(ix) Sugarcane and all species of Saccharum;
(x) Tobacco and all species of Nicotiana;
(xi) Berseem and all species of Trifolium:
(xii) Sunflower and all species of Helianthus
(xiii) Wheat and all species of Triticum;
(xiv) Paddy and all species of Oryzae;
(xv) Cuttings, saplings and bud-woods of flowers or ornamental plants;
(xvi) Seeds and the plant material of fruits.

Every consignment of the articles herein before mentioned shall be accompanied with the official Phytosanitary Certificate issued by the authorized officer of the country of origin of consignment, containing additional declarations that they are free from specified pests.

4. Post-Entry Quarantine

Plants and seeds, which require Post-entry Quarantine, shall be grown in Post-entry Quarantine facilities, approved and certified by the Designated Inspection Authority, to conform to the conditions laid down by the Plant Protection Adviser. The period for which and the conditions under which, the plants and seeds shall be grown in such facilities shall be specified in the permit. The Post-entry Quarantine facilities shall be established and provided by the importer or his agent at his own cost and these shall be ready for use at the time of arrival of the consignment in India. The importer shall obtain a certificate from the Designated Inspection Authority who, after inspection of the Post-entry Quarantine facilities, shall certify that such Post-entry Quarantine facilities have been duly established and provided in accordance with the guidelines of the Plant Protection Adviser. The importer shall produce this certificate before the Officer-in-charge of the Quarantine Station at the entry point, at the time of arrival of the consignment.

(i) The Officer-in-charge of the Quarantine Station. If after inspection of the consignment is satisfied, shall accord quarantine clearance with Post-entry-quarantine condition on the production, by an importer, of a certificate from the Designated Inspection Authority. (ii) After according quarantine clearance with Post-entry Quarantine conditions to the consignments of plants and seeds requiring Post-entry Quarantine, the officer-in-charge of the Quarantine Station at the entry point shall inform the Designated Inspection Authority, having jurisdiction over the Post-entry Quarantine facility of their arrival at the location where such plants would be grown by the importer.

The importer shall inform in advance the Designated Inspection Authority having jurisdiction, about the time of planting of such materials.

The importer shall permit to the Designated Inspection Authority complete access to the Post-entry Quarantine facility for the inspection of plants and shall, at all times, abide by his instructions concerning the plants in the Post-entry Quarantine.

The Designated Inspection Authority shall inspect the plants grown in the Post-entry Quarantine facility of the importer for the detection of the pests and diseases.

(i) The Designated Inspection Authority shall permit the release of plants from Post-entry Quarantine, if they are found to be free from pests and diseases for the period specified in the permit for importation.

(ii) Where the plants in the Post-entry Quarantine are found to be affected by pests and diseases during the specified period:

a. the Designated Inspection Authority shall order the destruction or return to the country of origin of the affected consignment of whole or a part of the plant population in the Post-entry Quarantine if the pest or disease is exotic and

b. The Designated Inspection Authority shall advise the importer about the curative measures to be taken to the extent necessary, if the pest or disease is not exotic and permit the release of the affected population from the Post-entry Quarantine only after quarantine measures have been observed. Otherwise, the plants shall be ordered to be destroyed.

(iii) Where destruction of any plant population is ordered by the Designated Inspection Authority, the importer shall destroy the same in the prescribed manner under the supervision of Designated Inspection Authority.

(iv) 12. The importer shall be liable to pay the prescribed fee for inspection of plants in the Post-entry Quarantine facility as laid down in Schedule III.

5. Appeal and Revision

If an importer is aggrieved by the decision of a Designated Inspection Authority regarding the destruction of any plant population, he may appeal to the Plant

Protection Adviser within 7 days from the date of communication of the decision giving the grounds of appeal. It shall lawful for the Plant Protection Adviser to relay on the observation of the incidence of pests and diseases and Designated Inspection Authority and such other expert opinion, as he may deem necessary, for deciding the appeal[Chapter 5].

The memorandum of appeal shall set out the grounds on which the decision is challenged and shall accompany with a fee of Rs. 10.00 by a Treasury Challan. The Plant Protection Adviser may, at any time, call for the records relating to any case pending before the Designated Inspecting Authority for the purpose of satisfying itself as to the legality or propriety of any decision passed by that authority and may pass such order in relating thereto, as it thinks fit.

No such order shall be passed after the expiry of three months from the date of the decision: Moreover, the Plant Protection Adviser shall not pass any order prejudicial to any person, without giving him a reasonable opportunity of hearing.

6. Powers of Relaxation

The Central Government may, in public interest, relax any of the conditions of this order relating to the permit and the phyto-sanitary certificate in relating to the import of any consignment [Chapter 6].

Summary

The seed order, 1989 was approved by the government of India, Ministry of Agriculture (Department of Agriculture and Cooperation) on 27th October, 1989. This order suppresses Plants, Fruits and Seeds (Regulation of import into India) Order 1984. This order may be called the Plants, Fruits and Seeds (Regulation of Import into India) Order, 1989. This order consists of six chapters. Chapter 1 deals with title, scope and definitions of various terms. Chapter 2 deals with general conditions of import. Chapter 3 explains special conditions of import. Chapter 4 deals with post entry quarantine, Chapter 5 deals with appeals and revision and chapter 6 explains powers of relaxation.

Questions

1. **Explain title and scope of seed control order, 1989.**

2. **Describe briefly general conditions of import of seed, fruits and plants for purpose of consumption.**

3. **Discuss briefly general conditions of import of seed, fruits and plants for planting purpose.**

4. **Explain in brief the procedure of issuing permit for import of seeds, fruit and plants.**

5. **List various seed, fruits and plants that are permitted for import.**

6. **Discuss briefly procedure of post entry quarantine of seeds, fruits and plants used for planting purpose.**

7. **Give a brief account of procedure of appeal related to import of seed, fruit and plants.**

8. **Write short notes on the following:**

   (i) Inspection authority

   (ii) Plant protection advisor

   (iii) Power of relaxation

   (iv) Seed import

Chapter 8

National Seeds Policy, 2002

Introduction

India has achieved self sufficiency in the production of food grains over the past 50 years. During this period the production has increased from 50 million tonnes to over 200 million tonnes. Thus, the country has progressed from a situation of food shortages and imports to one of surpluses and exports. Having achieved food sufficiency, the aim now is to achieve food and nutritional security at the household level.

Seed is the most important input of agricultural production on which the efficacy of other agriculture inputs is dependent. Seeds of high quality are required to meet the demand of diverse agro-climatic conditions and intensive cropping systems. Sustainable increase in agriculture production and productivity is dependent, to a large extent, on development of new and improved varieties of crops and an efficient system for timely supply of quality seeds to farmers. The seed sector has made remarkable progress over the last three decades. The area under certified seeds has increased from less than 500 hectares in 1962-63 to over 5 lakh hectares in 1999-2000. The quantum of quality seeds has crossed 100 lakh quintals.

Aims and Objectives

The Seeds Act, 1966, Seeds Control Order, 1984 and the New Policy on Seeds Development, 1988, form the basis of promotion and regulation of the Seed Industry. In India, enactment of these seed legislations have made significant changes in economic and agricultural scenario. The main objectives of the National Seeds Policy are briefly presented as follows:

1. **To Enhance Seed Replacement Rate:** It is necessary to enhance the seed replacement rates of various crops for achieving future targets of the food production.
2. **To Provide Quality Seed at Cheaper rate:** There is need to increase in the production of quality seeds, in which the private sector is expected to play a major role. Both private and Public Sector Seed Organizations at both Central and State levels, have to adopt economic pricing policies which would seek to realize the true cost of production.
3. **Rapid Development of Seed Industry:** National Seed policy will create favorable climate for growth of a competitive and localized seed industry, The National Seeds Policy provides an appropriate climate for the seed industry to utilize available and prospective opportunities
4. **To encourage import of useful germplasm:** The National Seed policy will encourage import of useful germplasm that can be used for developing high yielding varieties and hybrids in different field crops.
5. **Conservation of Agro-biodiversity:** The National Seed policy will safeguard the interests of Indian farmers and the conservation of agro-biodiversity.
6. **Safety from GMO to human and animal health:** The National Seed policy emphasizes on the safe use of genetically modified organisms avoiding possible harm to human and animal health.
7. **Use of Biotechnology for varietal improvement:** The National Seed policy provides a conducive atmosphere for application of frontier sciences in varietal development and for enhanced investments in research and development.

Globalization and economic liberalization have opened up new opportunities as well as challenges.

Thrust Areas

There are several thrust areas of National Seeds Policy, 2002. Important thrust areas include (1) varietal development, (2) plant variety protection, (3) seed production, (4) quality assurance, (5) seed distribution and marketing, (6) infrastructure facilities, (7) transgenic plant varieties, (8) import of seeds and planting material, (9) export of seed, (10) promotion of domestic seed industry and (11) strengthening of monitoring system. These are briefly discussed hereunder.

1. Varietal Development

The development of new and improved varieties of plants and availability of such varieties to Indian farmers is of crucial importance for a sustained increase in agricultural productivity. Appropriate policy framework and programmatic interventions will be adopted to stimulate varietal development in tune with market trends, scientific-technological advances, suitability for biotic and abiotic stresses, location specific adaptability and farmers' needs.

2. Plant Variety Protection

An effective *sui generis* system for intellectual property protection will be implemented to stimulate investment in research and development of new plant varieties and to facilitate the growth of the Seed Industry in the country.

A Plant Varieties and Farmers' Rights Protection (PVP) Authority will be established which will undertake registration of extant and new plant varieties through the Plant Varieties Registry on the basis of varietal characteristics.

(i) **Criteria for Registration**: The registration of new plant varieties by the PVP Authority will be based on the criteria of novelty, distinctiveness, uniformity and stability.

(ii) The criteria of distinctiveness, uniformity and stability could be relaxed for registration of extant varieties, which will be done within a specified period to be decided by the PVP Authority.

(iii) **Registration:** Registration of all plant genera or species as notified by the Authority will be done in a phased manner.

(iv) **Characterization:** The PVP Authority will develop characterization and documentation of plant varieties registered under the PVP Act and cataloguing facilities for all varieties of plants.

(v) **Farmers' Rights:** The rights of farmers to save, use, exchange, share or sell farm produce of all varieties will be protected, with the proviso that farmers shall not be entitled to sell branded seed of a protected variety under the brand name.

(vi) **Researchers' Rights**: The rights of researchers to use the seed/planting material of protected varieties for bona-fide research and breeding of new plant varieties will be ensured.

(vii) **Benefit Sharing:** Equitable sharing of benefit arising out of the use of plant genetic resources that may accrue to a breeder from commercialization of seeds/planting materials of a new variety, will be provided.

(viii) **Rewards:** Farmers/groups of farmers/village communities will be rewarded suitably for their significant contribution in evolution of a plant variety subject to registration. The contribution of traditional knowledge in agriculture needs to be highlighted through suitable mechanisms and incentives.

(ix) **Gene Fund:** A National Gene Fund will be established for implementation of the benefit sharing arrangement, and payment of compensation to village communities for their contribution to the development and conservation of plant genetic resources and also to promote conservation and sustainable use of genetic resources. Suitable systems will be worked out to identify the contributions from traditional knowledge and heritage.

(x) **Access to Germplasm**: Plant Genetic Resources for Food and Agriculture Crops will be permitted to be accessed by Research Organizations and Seed Companies from public collections as per the provisions of the

'Material Transfer Agreement' of the International Treaty on Plant Genetic Resources and the Biological Diversity Bill.

(xi) **Interaction between Public and Private Sector**: Regular interaction amongst the Private and Public Researchers, Seed Companies/ Organizations and Development Agencies will be fostered to develop and promote growth of a healthy seed industry in the country.

(xii) **Institutional Linkages:** To keep abreast of global developments in the field of Plant Variety Protection and for technical collaboration, India may consider joining Regional and International Organizations.

(xiii) **Compulsory Licensing:** The PVP Authority may, if required, resort to compulsory licensing of a protected variety in public interest on the ground that requirements of the farming community for seeds and propagating material of a variety are not being met or that the production of the seeds or planting material of the protected variety is not being facilitated to the fullest possible extent.

3. Seed Production

To meet the Nation's food security needs, it is important to make available to Indian farmers a wide range of seeds of superior quality, in adequate quantity on a timely basis. Public Sector Seed Institutions will be encouraged to enhance production of seed towards meeting the objective of food and nutritional security.

(i) **Categories of Seed:** The Indian seed program adheres to the limited three generation system of seed multiplication, namely, breeder, foundation and certified seed.

(ii) **Nucleus Seed:** It is the seed produced by the breeder to develop the particular variety and is directly used for multiplication as breeder seed.

(iii) **Breeder Seed:** Breeder seed is the progeny of nucleus seed.Breeder seed is the seed material directly controlled by the originating or the sponsoring breeder or Institution for the initial and recurring production of foundation seed.

(iv) **Foundation Seed:** It is the progeny of breeder seed. Foundation seed may also be produced from foundation seed. Production of foundation seed stage-I and stage-II may thus be permitted, if supervised and approved by the Certification Agency and if the production process is so handled as to maintain specific genetic purity and identity.

(v) **Certified seed:** It is the progeny of foundation seed or the progeny of certified seed. If the certified seed is the progeny of certified seed, then this reproduction will not exceed three generations beyond foundation stage-I and it will be ascertained by the Certification Agency that genetic identity and genetic purity has not been significantly altered.

(vi) **Access to Breeder Seed:** Public Sector Seed Production Agencies will continue to have free access to breeder seed under the National Agriculture Research System. The State Farms Corporation of India and National Seeds

Corporation will be restructured to make productive use of these organizations in the planned growth of the Seed Sector. Private Seed Production Agencies will also have access to breeder seed subject to terms and conditions to be decided by Government of India.

(vii) **Responsibility of Seed Production:** State Agriculture Universities/ICAR Institutes will have the primary responsibility for production of breeder seed as per the requirements of the respective States.

(viii) **Seed Quality:** Special attention will be given to the need to upgrade the quality of farmers' saved seeds through interventions such as the Seed Village Scheme.

(ix) **Rate of Seed Replacement:** Seed replacement rates will be raised progressively with the objective of expanding the use of quality seeds.

(x) **National Seed Map:** DAC, in consultation with ICAR and States, will prepare a National Seed Map to identify potential, alternative and non-traditional areas for seed production of specific crops.

(xi) **Perspective Plan:** To put in place an effective seed production program, each State will undertake advance planning and prepare a perspective plan for seed production and distribution over a rolling (five to six year) period. Seed Banks will be set up in non-traditional areas to meet demands for seeds during natural calamities.

(xii) **Seed Village Scheme:** The 'Seed Village Scheme' will be promoted to facilitate production and timely availability of seed of desired crops/varieties at the local level. Special emphasis will be given to seed multiplication for building adequate stocks of certified/quality seeds by providing foundation seed to farmers.

(xiii) **Popularizing New Varieties:** For popularizing newly developed varieties and promoting seed production of these varieties, seed mini-kits of pioneering seed varieties will be supplied to farmers. Seed exchange among farmers and seed producers will be encouraged to popularize new/non-traditional varieties. Seeds of newly developed varieties must be made available to farmers with minimum time gap. Seed producing agencies will be encouraged to tie up with Research Institutions for popularization and commercialization of these varieties.

(xiv) **Support for Hybrid Seed Production**: As hybrids have the potential to improve plant vigor and increase yield, support for production of hybrid seed will be provided.

(xv) **Seed Production in New Areas:** Seed production will be extended to agro-climatic zones which are outside the traditional seed growing areas, in order to avoid non-remunerative seed farming in unsuitable areas.

(xvi) **Establishment of Seed Banks:** Seed Banks will be established for stocking specified quantities of seed of required crops/varieties for ensuring timely and adequate supply of seeds to farmers during adverse situations such as natural calamities, shortfalls in production, etc. Seed Banks will be suitably strengthened with cold storage and pest control facilities.

(xvii) **Seed Storage:** The storage of seed at the village level will be encouraged to facilitate immediate availability of seeds in the event of natural calamities and unforeseen situations. For the storage of seeds at farm level, scientific storage structures will be popularized and techniques of scientific storage of seeds will be promoted among farmers as an extension practice.

(xviii) **Seed Crop Insurance:** Seed growers will be encouraged to avail of Seed Crop Insurance to cover risk factors involved in production of seeds. The Seed Crop Insurance Scheme will be reviewed so as to provide effective risk cover to seed producers and will be extended to all traditional and non-traditional areas covered under the seed production programme.

4. Quality Assurance

The Seeds Act will be revised to regulate the sale, import and export of seeds and planting materials of agriculture crops including fodder, green manure and horticulture and supply of quality seeds and planting materials to farmers throughout the country.

(i) **Establishment of National Seeds Board:** The National Seeds Board (NSB) will be established in place of existing Central Seed Committee and Central Seed Certification Board. The NSB will have permanent existence with the responsibility of executing and implementing the provisions of the Seeds Act and advising the Government on all matters relating to seed planning and development. The NSB will function as the apex body in the seed sector.

(ii) **Registration:** All varieties, both domestic and imported varieties, that are placed on the market for sale and distribution of seeds and planting materials will be registered under the Seeds Act. However, for vegetable and ornamental crops a simple system of varietal registration based on "breeders declaration" will be adopted.

(iii) The Board will undertake registration of kinds/varieties of seeds that are to be offered for sale in the market, on the basis of identified parameters for establishing value for cultivation and usage (VCU) through testing/trialing.

(iv) **Period of Registration:** Registration of varieties will be granted for a fixed period on the basis of multi-location trials to determine VCU over a minimum period of three seasons, or as otherwise prescribed as in the case of long duration crops and horticultural crops. Samples of the material for registration will be sent to the NBPGR for retention in the National Gene Bank.

(v) Varieties that are in the market at the time of coming into force of the revised Seeds Act, will have to be registered within a fixed time period, and subjected to such testing as will be notified.

(vi) The NSB will accredit ICAR, SAUs, public/private organizations to conduct VCU trials of all varieties for the purpose of registration as per prescribed standards.

(vii) The NSB will maintain the National Seeds Register containing details of varieties that are registered. This will help the Board to coordinate and assist activities of the States in their efforts to provide quality seeds to farmers.

(viii) The NSB will prescribe minimum standards (of germination, genetic characteristics, physical purity, seed health, etc.) as well as suitable guidelines for registration of seed and planting materials.

(ix) .Provisional registration would be granted on the basis of information filed by the applicant relating to trials over one season to tide over the stipulation of testing over three seasons before the grant of registration.

(x) Government will have the right to exclude certain kinds or varieties from registration to protect public order or human, animal and plant life and health, or to avoid serious prejudice to the environment.

(xi) The NSB will have the power to cancel the registration granted to a variety if the registration has been obtained by misrepresentation or concealment of essential data, the variety is obsolete and has outlived its utility and if the prevention of commercial exploitation of such variety is necessary in the public interest.

(xii) Registration of Seed Processing Units will be required if such Units meet the prescribed minimum standards for processing the seed.

(xiii) Seed Certification will continue to be voluntary. The Certification tag/label will provide an assurance of quality to the farmer.

(xiv) The Board will accredit individuals or organizations to carry out seed certification including self-certification on fulfillment of criteria as prescribed.

(xv) To meet quality assurance requirements for export of seeds, Seed Testing facilities will be established in conformity with ISTA and OECD seed certification programs.

(xvi) The State Government, in conformity with guidelines and standards specified by the Board, will establish one or more State Seed Testing Laboratories or declare any Seed Testing Laboratory in the Government or non-Government Sector as a State Seed Testing Laboratory where analysis of seeds will be carried out in the prescribed manner.

(xvii) Farmers will be encouraged to use certified seeds to ensure improved performance and output.

(xviii) **Farmers' Rights:** Farmers will retain their right to save, use, exchange, share or sell their farm seeds and planting materials without any restriction. They will be free to sell their seed on their own premises or in the local market without any hindrance provided that the seed is not branded. Farmers' right to continue using the varieties of their choice will not be infringed by the system of compulsory registration.

(xix) **Strict Measures:** Stringent measures would be taken to ensure the availability of high quality of seeds and check the sale of spurious or misbranded seeds.

5. Seed Distribution and Marketing

The availability of high quality seeds to farmers through an improved distribution system and efficient marketing set-up will be ensured to facilitate greater security of seed supply.

(i) For promoting efficient and timely distribution and marketing of seed throughout the country, a supportive environment will be provided to encourage expansion of the role of the private seed sector. Efforts will be made to achieve better coordination between State Governments to facilitate free Inter-State movement of seed and planting material through exemption of duties and taxes.

(ii) Private Seed Sector will be encouraged and motivated to restructure and reorient their activities to cater to non-traditional areas.

(iii) A mechanism will be established for collection and dissemination of market intelligence regarding preference of consumers and farmers.

(iv) A National Seed Grid will be established as a data-base for monitoring of information on requirement of seed, its production, distribution and preference of farmers on a district-wise basis.

(v) Access to term finance from Commercial Banks will be facilitated for developing efficient seed distribution and marketing facilities for growth of the seed sector.

(vi) Distribution and marketing of seed of any variety, for the purpose of sowing and planting will be allowed only if the said variety has been registered by the National Seeds Board.

(vii) National Seeds Board can direct a dealer to sell or distribute seeds in a specified manner in a specified area if it is considered necessary to the public interest.

6. Infrastructure Facilities

To meet the enhanced requirement of quality/certified seeds, creation of new infrastructure facilities along with strengthening of existing facilities, will be promoted.

National Seed Research and Training Center will be set up to impart training and build a knowledge base in various disciplines of the seed sector.

(i) The Central Seed Testing Laboratory will be established at the National Seed Research and Training Center to perform referral and other functions as required under the Seeds Act.

(ii) Seed processing capacity will be augmented to meet the enhanced requirement of quality seed.

(iii) Modernization of seed processing facilities will be encouraged in terms of modern equipment and latest techniques, such as seed treatment for enhancement of performance of seed, etc.

(iv) Conditioned storage for breeder and foundation seed and aerated storage for certified seed would be created in different regions.

(v) A computerized National Seeds Grid will be established to provide information on availability of different varieties of seeds with production agencies, their location, quality etc. This network will facilitate optimum utilisation of available seeds in every region.

(vi) Initially, seed production agencies in the public sector would be connected with the National Seed Grid, but progressively the private sector will be encouraged to join the Grid for providing a clear assessment of demand and supply of seeds.

(vii) State Governments, or the National Seeds Board in consultation with the concerned State Government, may establish Seed Certification Agencies.

(viii) State Governments will establish appropriate systems for effective execution and implementation of the objectives and provisions of the Seeds Act.

7. Transgenic Plant Varieties

Biotechnology will play a vital role in the development of the agriculture sector. This technology can be used not only to develop new crops/varieties, which are tolerant to disease, pests and abiotic stresses, but also to improve productivity and nutritional quality of food.

(i) All genetically engineered crops/varieties will be tested for environment and bio-safety before their commercial release, as per the regulations and guidelines of the Environment Protection Act (EPA), 1986.

(ii) The EPA, 1986, read with the Rules, 1989 would adequately address the safety aspects of transgenic seeds/planting materials. A list will be generated from Indian experience of transgenic cultivars that could be rated as environmentally safe.

(iii) Seeds of transgenic plant varieties for research purposes will be imported only through the National Bureau of Plant Genetic Resources (NBPGR) as per the EPA, 1986.

(iv) Transgenic crops/varieties will be tested to determine their agronomic value for at least two seasons under the All India Coordinated Project Trials of ICAR, in coordination with the tests for environment and bio-safety clearance as per the EPA before any variety is commercially released in the market.

(v) After the transgenic plant variety is commercially released, its seed will be registered and marketed in the country as per the provisions of the Seeds Act.

(vi) After commercial release of a transgenic plant variety, its performance in the field,will be monitored for at least 3 to 5 years by the Ministry of Agriculture and State Departments of Agriculture.

(vii) Transgenic varieties can be protected under the PVP legislation in the same manner as non-transgenic varieties after their release for commercial cultivation.

(viii) All seeds imported into the country will be required to be accompanied by a certificate from the Competent Authority of the exporting country regarding their transgenic character or otherwise.

(ix) If the seed or planting material is a product of transgenic manipulation, it will be allowed to be imported only with the approval of the Genetic Engineering Approval Committee (GEAC), set up under the EPA, 1986.

(x) Packages containing transgenic seeds/planting materials, if and when placed on sale, will carry a label indicating their transgenic nature. The specific characteristics including the agronomic/yield benefits, names of the transgenes and any relevant information shall also be indicated on the label.

(xi) Emphasis will be placed on the development of infrastructure for the testing, identification and evaluation of transgenic planting materials in the country.

8. Import of Seed and Planting Material

The objective of the import policy is to provide the best planting material available anywhere in the world to Indian farmers, to increase productivity, farm income and export earnings, while ensuring that there is no deleterious effect on environment, health and bio-safety.

(i) While importing seeds and planting material, care will be taken to ensure that there is absolutely no compromise on the requirements under prevailing plant quarantine procedures, so as to prevent entry into the country of exotic pests, diseases and weeds detrimental to Indian agriculture.

(ii) All imports of seeds will require a permit granted by the Plant Protection Advisor to the Government of India, which will be issued within the minimum possible time frame.

(iii) All import of seeds and planting materials, etc. will be allowed freely subject to EXIM Policy guidelines and the requirements of the Plants, Fruits and Seeds (Regulation of import into India) Order, 1989 as amended from time to time. Import of parental lines of newly developed varieties will also be encouraged.

(iv) Seeds and planting materials imported for sale into the country will have to meet minimum seed standards of seed health, germination, genetic and physical purity as prescribed.

(v) All importers will make available a small sample of the imported seed to the Gene Bank maintained by NBPGR.

(vi) The existing policy, which permits free import of seeds of vegetables, flowers and ornamental plants, cuttings, saplings of flowers, tubers and bulbs of flowers by certain specified categories of importers will continue. Tubers and bulbs of flowers will be subjected to post-entry quarantine.

(vii) After the arrival of consignments at the port of entry, quarantine checks would be undertaken; which may include visual inspection, laboratory inspection, fumigation and grow-out tests. For the purpose of these checks, samples will be drawn and the tests will be conducted concurrently.

9. Export of Seed

Due to diversity of agro-climatic conditions, strong seed production infrastructure and market opportunities, India holds significant promise for export of seeds.

(i) Government will evolve a long term policy for export of seeds with a view to raise India's share of global seed export from the present level of less than 1 per cent to 10 per cent by the year 2020.

(ii) The export policy will specifically encourage custom production of seeds for export and will be based on long term perspective, dispensing with case to case consideration of proposals.

(iii) Establishment and strengthening of Seeds Export Promotion Zones with special incentives from the Government will be facilitated.

(iv) A data bank will be created to provide information on the International Market and on export potential of Indian varieties in different parts of the world.

(v) A data base on availability of seeds of different crops to assess impact of exports on domestic availability of seeds will be created.

(vi) Promotional programmes to improve the quality of Indian seeds to enhance its acceptability in the International Market will be taken up.

(vii) Testing and certification facilities will be established in conformity with international requirements.

10. Promotion of Domestic Seed Industry

Incentives will be provided to the domestic seed industry to enable it to produce seeds of high yielding varieties and hybrid seeds at a faster pace to meet the challenges of domestic requirements.

(i) Seed Industry will be provided with a congenial and liberalized climate for increasing seed production and marketing, both domestic and international.

(ii) Membership to International Organisations and Seed Associations like ISTA, OECD, UPOV, ASSINSEL, WIPO, at the National level or at the level of individual seed producing agencies, will be encouraged.

(iii) Emphasis will be given to improving the quality of seed produced and special efforts will be directed towards improving the quality of farmers' saved seeds.

(iv) Financial support for capital investment, working capital and infrastructure strengthening will be facilitated through NABARD/Commercial Banks/ Cooperative Banks.

(v) Tax rebate/concessions will be considered on the expenditure incurred on in-house research and development of new varieties and other seed related research aspects. In order to develop a competitive seed market, the States will be encouraged to remove unnecessary local taxation on sales of seeds.

(vi) To encourage seed production in non-traditional areas including backward areas, special incentives such as transport subsidy will be provided to seed producing agencies operating in these marginalised areas.

(vii) Reduction of import duty will be considered on machines and equipment used for seed production and processing which are otherwise not manufactured in the country.

11. Strengthening of Monitoring System

The Department of Agriculture and Cooperation (DAC) will supervise the overall implementation and monitoring of the National Seeds Policy.

(i) The physical infrastructure in terms of office automation, communication facilities, etc., in DAC will be augmented in a time bound manner.

(ii) The technical capacity of DAC need to be augmented and strengthened to undertake the additional work relating to implementation of National Seeds Policy, implementation of PVP and FR Bill, Seeds Act, Import and Export of Seeds, etc.

(iii) Capacity building, including National and International training and participation in Seminars/Workshops will be organized for concerned officials.

National Seed Policy, 2009

Main intent of the National Seeds Policy is to make available and promote utilization of best quality seeds and planting material of diverse elite genetic resources; to enhance and sustain crop productivity, improve food security, generate additional sources of farm income and employment; provide base for value-added agricultural exports and alleviate poverty in rural areas of Pakistan.

Enabling environment will be created for participation of private sector in variety development, variety evaluation for VCU and DUS examination, early generation pedigreed seed production, accredited seed quality testing, production of seed processing machinery and biotechnology research through appropriate incentives.

Plant Breeder's Rights will be protected to stimulate investment in research and development of new plant varieties and to facilitate the import of elite genetic resources and hybrid seed production technologies.

Biotechnology research will be strengthened to accelerate development of useful gene constructs for incorporation in plant breeding programs.

Integration of seed market at national level will be achieved, in consultation with the Provincial Seed Councils, to pave the way for nation-wide variety release on the basis of national variety performance trials.

Seed exports will be promoted through participation in regional seed schemes; OECD seed scheme and other initiatives under SAARC, ECO or bilateral arrangements.

Summary

The Government of India trusts that the National Seeds Policy will receive the fullest support of State Governments/Union Territory Administrations, State Agricultural Universities, plant breeders, seed producers, the seed industry and all other stakeholders, so that it may serve as a catalyst to meet the objectives of sustainable development of agriculture, food and nutritional security for the population, and improved standards of living for farming communities.

The National Seeds Policy will be a vital instrument in attaining the objectives of doubling food production and making India hunger free. It is expected to provide the impetus for a new revolution in Indian agriculture, based on an efficient system for supply of seeds of the best quality to the cultivator.

The National Seeds Policy will lay the foundation for comprehensive reforms in the seed sector. Significant changes in the existing legislative framework will be effected accompanied by programmatic interventions. The Policy will also provide the parameters for the development of the seed sector in the Tenth and subsequent Plans. The progress of implementation of the Policy will be monitored by a High Level Review Committee.

Questions

1. **Describe briefly objectives of national Seeds Policy, 2002.**

2. **List thrust areas of national Seeds Policy, 2002. Describe briefly plant variety protection.**

3. **Discuss various aspects of seed production in relation to national Seeds Policy, 2002.**

4. **Describe various quality assurance measures outlined in National Seeds Policy, 2002.**

5. **Describe various guidelines of seed distribution and marketing in relation to national Seeds Policy, 2002.**

6. **Explain various infrastructure facilities suggested by national Seeds Policy, 2002.**

7. **Describe National Seeds Policy in relation to transgenic plant varieties.**

1. **What are the guidelines suggested for import of seed and planting material by national Seeds Policy, 2002?**

2. **Describe the guidelines suggested for export of seed material by national Seeds Policy, 2002.**

3. Explain various guidelines suggested for promotion of domestic seed Industry by national Seeds Policy, 2002.

4. What ways and means have been suggested for strengthening of monitoring system by national Seeds Policy, 2002?

5. Write short notes on the following:

(i) Farmers' Rights

(ii) Researchers' Rights

(iii) National Gene Fund

(iv) National Seeds Board

Chapter 9

Indian Seed Act 2004

Introduction

The seed can be defined in two ways, *viz.* in broad sense and in strict sense. In the broad sense, any plant part which is used for commercial multiplication of a crop is called seed. For example, in case of sugarcane and potato stem and tuber are also known as seed. In strict sense, seed is the product of fertilized ovule that consists of embryo, seed coat and cotyledon (s).The seed of a release and popular variety produced by scientific method is referred to as improved seed or quality seed. Variety refers to a genotype which has been released for commercial cultivation either by State Variety Release Committee or Central Variety Release Committee. Improved seed plays an important role in maximizing production and productivity of field crops. It results in (i) better germination, (ii) vigorous seedling growth, (iii) higher crop stand, (iv) better quality of produce, and (v) ultimately in higher crop yield.

Seed Act refers to the legal procedures approved by the government for production and marketing of seeds. The main objective of the Seeds Act is to ensure availability of quality seeds to farmers. In India, the Seed Act was first introduced in 1966 which is called as Seeds Act 1966. This Act was amended in 1972 which is known as the Seeds amended Act 1972. Subsequently, the seed act was thoroughly revised and the revised act was approved by the Government of India in September 2004. The revised Seeds Act is known as Seeds Act 2004 which came into force with effect from January 2005. It replaces the Seeds Act, 1966.

Salient Features

The main issues of Indian Seed Act 2004 in relation to coverage, seed category, license, registration, criteria for protection, period of protection, certification, compensation, farmers' privilege, penalty, exclusion, research exemption and constitution of central seed committee are briefly presented below:

1. **Coverage:** This Act has wider coverage than seeds act of 1966. It includes agriculture, horticulture, forestry, plantation crops, medicinal and aromatic plants. The seeds of all these crops are covered by the new act.
2. **Seed Category:** This act permits sale of certified seed only The name of variety, physical purity and germination percentage have to be indicated on the seed container or bag.
3. **License:** The seed dealers, sellers and growers should have license which is compulsory.
4. **Registration:** Registration for all kinds and varieties of seed is compulsory. No producer can grow seeds unless he is registered. It is compulsory for all cultivated varieties of field crops, vegetable crops, fruit crops, medicinal and aromatic plants, forest species and plantation crops. Every seed producer and dealer, and horticulture nursery has to be registered with the state government.
5. **Criteria for Registration:** The released and notified varieties can be registered for seed multiplication. There are four criteria for protection of varieties under PPV and FR Act. These are novelty, distinctiveness, uniformity and stability.
6. **Period of Registration:** The Indian Seeds Act 2004 has adopted UPOV Act 1978. According to this Act, the period of protection is 15 years for annual and biennial crops and 18 years for perennial plants such as trees and vines.
7. **Certification:** The Act permits self certification of seeds by accredited agencies and allows the central government to recognize certification by foreign seed certification agencies. Seed producers are permitted to self certify the performance of their seed under certain conditions. The seed companies need to provide the results of multiplication trials before registration.
8. **Compensation:** The disputes about the quality of seeds have to be settled in the consumer court. Any loss in the production due to poor seed quality is claimed in the consumer court.
9. **Farmers' Privilege:** The Seeds Act 2004 allows farmers' privilege. The Bill does not restrict the farmer's right to use or sell his farm seeds and planting material, provided he does not sell them under a brand name. All seeds and planting material sold by farmers will have to conform to the minimum standards of germination, physical purity and genetic purity applicable to registered seeds.
10. **Penalty:** Any person who contravenes any provisions of the Act or imports, sells or stocks seeds deemed to be misbranded or not registered, can be punishable by a fine between Rs. 5,000 and Rs. 25,000. The penalty for

giving false information is a prison term up to six months and/or a fine up to Rs 50,000. In case of companies, the person(s) in charge of conduct of the business of the company will be held accountable.

11. **Exclusion:** Those techniques which are harmful to animals and the environment such as terminator technology and traitor technology are excluded and not allowed by the seed act.
12. **Research Exemption:** This act allows use of protected material by breeders for development of new varieties.
13. **Central Seed Committee:** In the Seed Act 2004, the central seed committee consists of Chairman [Secretary DARE], Agricultural Commissioner, DDG horticulture [ICAR], Joint Secretary Seeds[DAC], representatives from DBT, Ministry of Environment and Forest, Secretary Agriculture [from five States], Director [SSCA] from one State and MD-SSC from one State. In addition, two representatives each of farmers and seed industry are nominated.

Challenges

1. **Reduction in Biodiversity**: The new seeds act will encourage cultivation of new varieties. Thus old varieties and land races will be replaced by new varieties on vast areas resulting in significant reduction in biodiversity. Reduction in biodiversity will invite danger of uniformity and lead to narrow genetic base and narrow adaptation.
2. **Monopoly of Seed Companies**: It will encourage multinational seed companies which may lead to monopoly of such companies. Seed companies may hike seed prices resulting in exploitation of farmers.
3. **Difficult to Trace Defective Seed Lot:** The Seed Act does not provide for a mechanism to trace back a packet of seed to the dealer, processor and producer. This would make it difficult to trace back a defective lot, and rectify any deficiencies in the supply chain.
4. **False Certification:** Seed producers would be permitted to self-certify the performance of their seeds under certain conditions. This opens up the possibility of false declaration by seed companies. To prevent this, only government agencies should be allowed to conduct these trials and grant certification.
5. **Maintenance of Records:** Every horticultural nursery has to be registered with the state government and has to maintain records of layout plan, source of every planting material etc. Such nurseries in the unorganized sector may find it difficult to adhere to these conditions.

Comparison of Seed Act 1966 and Seed Act 2004

There are some similarities and some dissimilarities between two seeds acts. A brief comparison of seeds act 1966 and seeds act 2004 is presented in Table 9.1.

Table 9.1: Comparison of Seed Act 1966 and Seed Act 2004

Sl.No.	*Particulars*	*Seeds Act 1966*	*Seeds Act 2004*
A.	**Similarities**		
1.	Category of seed permitted for sale	Certified Seed	Certified Seed
2.	License	Compulsory	Compulsory
3.	Certification	Compulsory	Compulsory
4.	Farmers' exemptions	Available	Available
5.	Purity and germination standards	Adopted	Adopted
6.	Compensation of loss by use of certified seed	Through consumer court	Through consumer court
7.	Requirements for seed multiplication	Released and notified varieties	All seeds for sale must be registered
B.	**Dissimilarities**		
1.	Crop covered	Agriculture and horticulture	Agriculture, horticulture, forestry, plantation crops, medicinal and aromatic plants
2.	Registration of transgenic varieties	No provision	Special provision
3.	Registration with PPVFR authority	Not required	Required
4.	Period of protection	Not defined	Defined
5.	Penalty for violation of Act	Rs 100-1000/and 6 months imprisonment	Rs 500-50,000/and six months imprisonment
6.	Self Certification	Not permitted	Permitted
7.	Seed multiplication of land Races	Permitted	Not permitted
8.	Representatives in central seed committee	From all States	From five States only
9.	Involvement of Private Sector	No	Yes

Comparison of Seed Act 2004 and PPV and FR Act 2001

There are some similarities and some dissimilarities between Seed Act 2004 and Protection of Plant Variety and Farmers' Rights Act 2001 which are presented in Table 9.2.

Merits of Seed Act 2004

1. **Wider Coverage:** This act has wider coverage [agriculture, horticulture, forestry, plantation crops, medicinal and aromatic plants] than Seed Act of 1966.
2. **Easy Certification:** This act has made the process of certification very simple means allowed self certification. But it should not be misused.
3. **Bans Harmful Technologies:** The act did not permit registration of harmful technologies such as terminator technology and traitor technology.

Table 9.2: Comparison of Seeds Act 2004 and Protection of Plant Varieties and Farmers' Rights Act 2001

Sl.No.	*Particulars*	*Seeds Act 2004*	*PPV and FR Act 2001*
A.	**Similarities**		
1.	Registration	Compulsory	Compulsory
2.	Period of registration	15 years for annual and biennial crops and 18 years for perennials	15 years for annual and biennial crops, 18 years for perennials
3.	Renewal of registration	Permitted for similar period	Permitted for similar period
4.	Farmers Rights	Allowed	Allowed
5.	Plant breeders' Rights	Available	Available
6.	Disclosure of variety's performance	Compulsory	Compulsory
7.	Sale of branded seeds by farmers	Not permitted	Not permitted
B.	**Dissimilarities**		
1.	Record Maintained	National Register of Seeds	National register of plant varieties
2.	Compensation	Through consumer court	Through PPV and FR Authority
3.	Disclosure of variety's parentage	Not required	Required
4.	Variety's performance to be declared by	Seed producer, distributor or vendor	Breeder of the variety
5.	Penalty for violation of the Act	Rs 5000-50,000/	Rs. 50,000-20lakh
6.	Imprisonment for violation of Act	Up to six months	From 2-3 years

4. **Permits Farmers' Rights:** It permits farmers' rights. The farmers need not to register their varieties.
5. **Ensures Quality Seed:** This act ensures availability of good quality seeds to farmers by imposing heavy penalties on the sale of spurious seeds.
6. **Registration for long Duration:** This act provides registration for 15 years for annual and biennial crops and 18 years for perennial crops which is sufficiently long period. Moreover, the registration is renewable for another term of similar duration.
7. **Central Seed Committee:** In the Seed Act 2004, the constitution of central seed committee is better than that of Seed Act 1966.
8. **Promotes Seed Industry:** The new seed act will lead to fast development of seed industry due to involvement of representatives of multinational seed companies or industries in the central seed committee.
9. **Increase in Productivity:** The new act will help in increasing crop productivity due to availability of better quality seeds.
10. The present seed replacement rate is 20-25 per cent. The seed act will enhance seed replacement rate of various crops.

11. The seed act will boost the export of seeds and encourage import of useful germplasm for use in crop improvement programmes.
12. It will encourage investment in seed research and development.

Demerits of Seed Act 2004

1. Although farmers are exempt from registering their seed varieties, the seeds have to conform to standards prescribed for commercial seeds. Farmers may find it difficult to adhere to the standards required of commercially sold seeds, because the seed saved and exchanged by farmers constitute above 80 per cent of the seeds planted.
2. It is not clear whether the compensation would include the value of the crop or only the cost of the seed.
3. It is not clear whether a seed producer may sell seed which is registered by a different producer.
4. The disclosure of parentage is not essential. It may lead to a situation where seeds can be registered without disclosing the parentage or origin of the seed.
5. The Act does not have the provision of benefit sharing unlike the Convention on Biological Diversity and the PPVFR Act.
6. The Seed Inspector has the power to enter and search as well as break open container or break open doors, without a warrant. This is against the normal procedure.

Summary

Seed Act refers to the legal procedures approved by the government for production and marketing of seeds. The main objective of the Seeds Act is to ensure availability of quality seeds to farmers. In India, the Seed Act was first introduced in 1966 which is called as Seeds Act 1966. This Act was amended in 1972 which is known as the Seeds amended Act 1972. Subsequently, the seed act was thoroughly revised and the revised act was approved by the Government of India in September 2004. The revised Seeds Act is known as Seeds Act 2004 which came into force with effect from January 2005. It replaces the Seeds Act, 1966. The significant milestones in the history of Indian Seed Legislation have been highlighted.

The new seed Act of 2004 has wider coverage than seeds act of 1966. It includes agriculture, horticulture, forestry, plantation crops, medicinal and aromatic plants. Main features, merits and demerits of Seed Act 2004 are presented. A comparison of Seeds Act 2004 is presented with Seeds Act 1966 and PPV and FR Act 2001. The future challenges have been outlined.

Questions

1. **Explain in brief main features of Indian Seeds Act 2004.**

2. **What are merits and demerits of Indian Seeds Act 2004?**

3. **Give a comparison of Indian Seeds Act 1966 and Seed Act 2004 in a tabular form.**

4. **Compare Indian Seeds Act 2004 and Protection of Plant Varieties and Farmers' Act 2001.**

5. **Describe briefly main features of Indian Seeds Act 1966.**

6. **Discuss briefly the advantages and limitations of Indian Seeds Act 1966.**

7. **Describe the constitution of Central Seed Committee of (i) Indian Seeds Act 1966 and (ii) Indian Seeds Act 2004.**

8. **What do the following terms signify?**
   (i) Extant Variety
   (ii) Transgenic variety
   (iii) Spurious Seed
   (iv) Farmers Rights

9. **Write short notes on the following:**
   (i) Farmers' Rights
   (ii) Farmers' Compensation
   (iii) Orthodox seeds
   (iv) Recalcitrant seeds

10. **Describe briefly India Seeds Act 2004 on the basis of following:**
    (i) Seed certification
    (ii) Registration
    (iii) Licensing
    (iv) Penalties

11. **Describe briefly the Structure of Central Seed Committee of Indian Seed Act 2004.**

12. **What are the similarities and dissimilarities between Seeds Act 1966 and Seeds Act 2004?**

13. **Describe briefly the differences between Indian Seeds Act 2004 and Protection of Plant varieties and farmers' Rights Act 2001**

14. **Write Short Notes on the following:**
    (i) Indian Seeds Act 1966
    (ii) Indian Seeds Act 2004
    (iii) PPV and FR Act 2001
    (iv) Biodiversity Act 2002

Section II

Varieties Related Legislations

Chapter 10

Protection of Plant Varieties and Farmers' Rights Act 2001

Introduction

The Protection of Plant Varieties and Farmers' Rights Act 2001 was approved by the government of India in 2004 but came into force with effect from. January 2005. The main objective of this act is to encourage development of new plant varieties and promote development of seed industry in India.

Main Features

The main features of PPV and FR Act 2001 in relation to purpose, legislative structure, headquarters, registration, period of registration, farmers' rights, plant breeders' rights, exclusion, compensation and penalties are briefly presented below.

1. **Purpose:** There are four main objectives of enforcing the Protection of Plant varieties and Farmers' Rights Act 2001 as given below.
   - **(i)** To establish an effective system for protection of plant varieties, farmers' rights and plant breeders' rights.
   - **(ii)** To recognize and protect farmers' rights for their contribution in conserving, improving and making available plant genetic resources for development of new plant varieties.
   - **(iii)** To stimulate investment for research and development both public and private sector for development of new plant varieties.
   - **(iv)** To promote the growth of seed industry in the country to ensure availability of high quality seeds to the farmers.

2. **Legislative Structure:** The Authority consists of a chair person called as authority and 15 persons as members. The members of the Authority include (i) Agricultural Commissioner-GOI, (ii) DDG Crop Sciences-ICAR, (iii) Joint Secretary Seeds-GOI, (iv) Horticulture Commissioner-GOI, (v) Director, NBPGR; (vi) one member each [not below the rank of Joint Secretary] from DBT, (vii) Ministry of Environment and forests-GOI, (viii) Ministry of Law Justice and Company Affairs- GOI; (ix) one representative each [nominated by central government] from farmers, (x) tribal organization, (xi) seed industry, (xii) an agricultural university, (xiii) and women organization; and (xi**v**) two nominated representative of state government.
3. **Headquarters:** The headquarters of Protection of Plant Varieties and Farmers' Rights Authority is located in Pusa Campus [National Agricultural Science Complex], New Delhi.
4. **Plant Material Protected:** Various types of crop cultivars can be protected under PPV and FR Act. The following breeding material can be protected.
   (i) All new varieties of self pollinated species such pure line and multi line varieties.
   (ii) New varieties of cross pollinated species such as open pollinated variety, synthetics and composites.
   (iii) New varieties of vegetatively/clonally propagated species.
   (iv) Single, three-way and double cross hybrids.
   (v) Inbred parental lines of hybrid varieties.
   (vi) Trees, fruits, ornamental plants and vines, etc.
5. **Registration:** The PPV and Fr Act Authority establishes a Plant Varieties Registry, which maintains a National Register of Plant Varieties. Specifies details under which a variety may be registered such as a complete passport data of the parental lines from which a variety has been derived have to be furnished by the breeder. A new variety will be registered under the act if it conforms to the criteria of novelty, distinctiveness, uniformity and stability.
   (i) Novelty: It refers to newness of a variety. A variety which has not been grown for more than one year.prior the application for registration can be protected.
   (ii) Distinctiveness: The variety should be clearly distinguishable fro previously available varieties at least for one characteristics.
   (iii) Uniformity: It refers to same type of population of a variety. In cross pollinated species, the range of variation should be within the tolerable limit.
   (iv) Stability: It means that the variety should give stable performanceafter repeated reproduction or propagation. Moreover, a variety should have a designation.

   For registration of a variety, an application has to be filed with the Registrar, PPV and FR Act. No fee is charged if a variety is registered by a farmer.

6. **Period of Protection:** The PPV and FR Authority has accepted UPOV Act 1978. According to this Act, the period of protection is 15 years for annual and biennial crops and 18 years for perennial plants such as trees and vines. The registration is renewable after its expiry for a similar term of duration.
7. **Farmers' Rights:** A farmer is entitled to save, use, sow, resow, exchange, share or sell his farm produce including seed of a variety protected under the Act in the same manner before this Act came into force. He cannot sell branded seed of a variety protected under the Act.
8. **Plant Breeders' Rights:** A certificate of registration for a variety issued by this act shall confer an exclusive right on the breeder or his successor, his agent or licensee, to produce, sell, market, distribute, import or export seed of his variety. The breeder has right to authorize any other person for production and marketing of his variety.
9. **Exclusion:** Those techniques which are harmful to animals and the environment such as terminator technology and traitor technology are excluded and not allowed by the Protection of Plant Varieties and farmers' Rights Act 2001.
10. **Compensation:** If a breeder of a propagating material of a variety registered under the Act sells his product to a farmer, he has to disclose the expected performance under given conditions. If the propagating material fails to perform, the farmer can claim compensation in the prescribed manner before the Protection of Plant Varieties and Farmers' Rights Authority.
11. **Penalty:** Penalty for applying false denomination to a variety is imprisonment up to two years and/or a fine between Rs 50,000 and Rs five lakh. Penalty for falsely representing a variety as registered is imprisonment up to three years and/or a fine between Rs one lakh and Rs five lakh or both. Penalty for subsequent offence is imprisonment up to three years and/or a fine between Rs two lakh and Rs 20 lakh.

Comparison of Seed Act 2004 and PPV and FR Act 2001

There are some similarities and some dissimilarities between Seed Act 2004 and Protection of Plant Variety and Farmers' Rights Act 2001 which are presented in Table 10.1.

Merits of PPV and FR Act 2001

It provides an effective system for protection of plant varieties, farmers' rights and plant breeders' rights:

1. It recognizes the contribution of farmers in conserving, improving and making available plant genetic resources for development of new plant varieties.
2. It provides rights to farmers to save, use, sow, resow, exchange, share or sell his farm produce including seed of a variety protected by PPV and FR Act.

Table 10.1: Comparison of Seed Act 2004 and Protection of Plant Varieties and Farmers' Rights Act 2001

Sl.No.	*Particulars*	*Seed Act 2004*	*PPV and FR Act 2001*
1.	Registration	Compulsory	Compulsory
2.	Period of registration	15 years for annual and biennial crops, 18 years for perennials	15 years for annual and biennial crops, 18 years for perennials
3.	Renewal of registration	Permitted for similar period	Permitted for similar period
4.	Farmers Rights	Allowed	Allowed
5.	Plant breeders' Rights	Available	Available
6.	Disclosure of variety's performance	Compulsory	Compulsory
7.	Sale of branded seeds by farmers	Not permitted	Not permitted
8.	Involvement of Private Sector	Yes	Yes
B.	**Dissimilarities**		
1.	Record maintained	National Register of Seeds	National register of plant varieties
2.	Compensation	Through consumer court	Through PPV and FR Authority
3.	Disclosure of variety's parentage	Not required	Required
4.	Variety's performance to be declared by	Seed producer, distributor or vendor	Breeder of the variety
5.	Penalty for violationof the Act	Rs. 5000-50,000/	Rs. 50,000-20lakh
6.	Imprisonment for violation of the Act	Up to six months	From 2-3 years

3. It confers an exclusive right on the breeder or his successor, his agent or licensee, to produce, sell, market, distribute, import or export seed of his varietyThus it gives recognition to breeders for their contribution and encourage them to breed still better varieties.
4. It encourages investment in plant breeding research and development both by public and private sectors for development of superior plant varieties.It will lead to fast development of seed industry and improvement in seed quality due to competitions among breeders.
5. It promotes the growth of seed industry in the country to ensure availability of high quality seeds to the farmers.
6. It involves representatives of farmers, women and seed industry in its legislative structure who will look to the interest of these sectors.
7. The Act has provision for compensation. If the seed gives poor performance than the declared by the breeder, farmer can claim compensation in the prescribed manner before the Protection of Plant Varieties and Farmers' Rights Authority.

8. The Act has provision for rewarding farmers for their role in conserving plant genetic resources. Such expenditure can be met out from the gene fund available with the PPV and FR Act Authority.
9. The act has provision of benefit sharing. The farmers can get share from the benefits arising from cultivation of farmers varieties and knowledge.
10. The act has provision to impose penalties to curb unhealthy practices.

Demerits of PPV and FR Act 2001

1. The PPV and FR Act will promote cultivation of new high yielding varieties which are highly uniform. This will result in reduction in genetic diversity leading to narrow genetic base and danger of uniformity.
2. This will encourage monopoly of some breeders or seed companies for genetic material with special traits.
3. The holders of plant breeders rights will produce less seed than market demand and hike price to earn more profit. The high price can not be afforded by small and marginal farmers Thus all farmers can not take advantage of new varieties.
4. It will inhibit free exchange of seed material and encourage unlawful practice.

Summary

The Protection of Plant Varieties and Farmers' Rights Act 2001 was approved by the government of India in 2004 but came into force with effect from. January 2005. The main objective of this act is to encourage development of new plant varieties and promote development of seed industry in India. The main features of PPV and FR Act 2001 in relation to purpose, legislative structure, headquarters, plant material protected, registration, period of registration, farmers' rights, plant breeders' rights, exclusion, compensation and penalties have been discussed.

There are four basic requirements for protection of a variety under PPV and FR Act, *viz.* (i) novelty, (ii) distinctiveness, (iii) uniformity, and (iv) stability. Varieties, hybrids and inbred lines of hybrids are protected under this act.

A comparison of Indian Seeds Act 2004 and Protection of Plant varieties and farmers' Rights Act 2001 has been given in tabular form. Merits and demerits of the act have been highlighted.

Questions

1. **Describe briefly main features of PPV and FR Act 2001**

2. **Discuss briefly merits and demerits of PPV and FR Act 2001**

3. **Compare Indian Seed Act 2004 and Protection of Plant Variety and Farmers' Act 2001.**

4. **Explain briefly legislative structure of PPV and FR Act 2001.**

5. **Write short notes on the following:**

   (i) Plant Breeders Rights **(ii)** Transgenic variety

   (iii) Farmers Right **(iv)** Candidate variety

6. **Describe briefly PPV and FR Act 2001 on the basis of following:**

   (i) Period of Protection **(ii)** Criteria for Registration

   (iii) Compensation **(iv)** Penalties

7. **Describe briefly the requirements for registration of a variety under PPV and FR Act 2001.**

8. **What do the following terms signify?**

   (i) Extant variety **(ii)** Reference variety

   (iii) Example variety **(iv)** Farmers'variety

9. **Write short notes on the following.**

   (i) Novelty **(ii)** Distinctiveness

   (iii) Uniformity **(iv)** Stability

10. **Give full form of the following**

   (i) PBR **(ii)** PVPA

   (iii) DUS **(iv)** UPOV

Chapter 11

Plant Breeders' Rights

Introduction

The Plant Breeders' Rights program was first established in 1987 under the Plant Variety Rights Act [PVRA], which was succeeded by the current Plant Breeders' Rights Act 1994. {PBRA].Plant breeders' rights [PBR], also known as plant variety rights (PVR), are intellectual property rights granted to the breeder of a new variety of plant. Plant Breeders Rights are granted to novel plant varieties that are distinctive, uniform, and stable (*e.g.*, cultivars breed true-to-type for desired traits). The legal protection of a new plant variety is granted to the breeder or his successor. The effect of PBR is that prior authorization is required before the material can be used for commercial purposes.

Main Features

Main features of Plant Breeders' Rights in relation to registration, duration, validity, matters covered, requirements, transfer, control, enforcement, etc. are briefly presented below:

1. **Registration:** Registration is essential to get legal rights. There is no protection without registration. The registration consists of following steps:
   - **(i)** Filing application in the prescribed form,
   - **(ii)** Payment of prescribed processing fee along with application. The registration fee varies from country to country. In USA, after registration annual renewal fee [300 Dollars] is payable.
   - **(iii)** Examination of the application by the experts, and
   - **(iv)** Issue of protection certificate, if application if found to meet the desired requirements.

(v) Normally the registration takes about 2.5 years for most of the species. takes 2.5 years.

2. **Duration:** The period of protection varies with plant species. The protection is provided for a definite period. The duration of protection for different plant species is as follows':

   (i) **For Trees and Vines:** Initially for nine years and maximum for 18 years.

   (ii) **For extant varieties:** Initially six years and maximum for 15 years fro the date registration of the variety.

   (iii) **In other cases**: Initially six years and maximum for 15 years fro the date registration of the variety.

   Thus breeders' rights allow plant breeders to get benefit of their variety for a limited period. When the PBR period is over, the variety reverts to the public domain and is available to every body. The limited duration of Plant Breeders' Rights ensure a balance between private and public interest.

3. **Validity:** The protection of the Plant Breeders' Rights is valid only in the country where it has been registered. The protection in other countries can be obtained by filling separate application in each country.

4. **Matters Covered:** The protection right can be granted for varieties of all botanical genera and species. The variety should have a designation [name] as per the rule of International Code of Nomenclature.

5. **Requirements:** There are four basic requirements for protection of a variety under PBR, *viz.* (i) novelty, (ii) distinctiveness, (iii) uniformity, and (iv) stability. See later.

6. **Transfer:** An authorized plant breeder has right to authorize other interested persons for commercial production and marketing of his variety on royalty basis.

7. **Control:** The authorized plant breeder has right to prevent other from commercial production and marketing of his variety without permission.

8. **Enforcement:** The Plant Breeders' Rights come into force immediately after registration of the variety.

9. **Other Points:** There are some other points which are taken into account for granting PBR. These are as follows:

   (i) BPR can be granted only to breeder or legal owner of a new variety.

   (ii) The plant breeders rights protect the variety but not the standard breeding procedures that are used to develop new variety.

   (iii) The main purpose of PBR is to encourage breeders to develop better plant varieties

   (iv) The PBRs provide monopoly of new varieties to breeders for a limited period.

   (v) The same variety can be patented and protected by PBR. The gene technology or gene sequences can be patented and the end product that is the plant variety can be protected under PBR.

(vi) A person growing protected variety can not sale its seed on commercial scale. But he can use the produce of such seed for planting purpose on his farm.

(vii) PBR coexists with other legislation. It is not independent of other laws of the land.

(viii) The protected variety can be used to breed new varieties.

(ix) A variety cannot be registered under trademarks. It comes under Sui generic rights, hence it can be registered under PBR only.

Requirements

The PBR cannot be granted to any variety. There are four basic requirements for protection of a variety under Plant Breeders' Rights. These are (i) novelty, (ii) distinctiveness, (iii) uniformity, and (iv) stability. These are defined as follows:

(i) Novelty: It refers to newness of a variety. The variety should be new and should not have been commercially cultivated for more than one year before granting protection under PBR. In other words, there should not be a prior sale for more than one year.

(ii) Distinctiveness: The new variety must be distinguishable in one or more morphological, quality or other character from previously available varieties. When a variety differs atleast in one characteristic from varieties of common knowledge, it is known as distinct.

(iii) Uniformity: The variety should be uniform. In other words, all plants in a variety should look alike.

(iv) Stability: The should give stable performance in different generations. All these attributes are determined by grow out test.

Rights Protected

The *Plant Breeders' Rights Act* provides plant breeders the exclusive right to produce and sell new plant varieties which they have developed. In other words, It provides exclusive rights to the breeder for commercial production and marketing of his variety. The important Plant Breeders' Rights are as follows:

1. **Rights for commercial Seed production**: PBR provide legal right to the breeder of a variety for large scale seed production. This he can do either on his own farm or on the farms of authorized farmers on payment basis.
2. **Rights for Marketing**: The breeder or owner of a variety has exclusive rights to regulate marketing of his variety.
3. **Rights to export and import**: The breeder or owner of a variety has full rights to regulate export and import of his or her variety.
4. **Rights of Authorization**: The breeder or owner of a protected variety has rights to authorize other interested persons for commercial production, and marketing, export and import of his variety. However, prior authorization of the breeder or owner of a variety is required for such purpose.

5. **Rights to prevent Infringement**: The breeder or owner of a variety has rights to prevent others from unauthorized commercial production and marketing of his variety.

Infringement

Unauthorized production and marketing of a registered variety by other person amounts to infringement. The owner has the right to take legal action against the infringer and claim damages. The PBR Authority may initiate legal action against a person who is involved in the infringement of a protected plant variety. The Authority can recover both damages and profits from such person. The Plant Breeders' Act [PBRA] provides for heavy penalties against infringement of the breeders' right. In USA it 55,00 Dollars for individuals and 275000 Dollars for companies.

Exemptions of PBRA

The Plant Breeders' Rights Act provides two exemptions, *viz*. (i) breeders' exemptions and (ii) farmers' exemptions. These are defined as follows:

(i) **Breeders' Exemptions:** The legal rights that are provided to plant breeders to use protected material for further research refer to breeders' exemptions. Breeders exemptions are also called research exemptions or breeders' privilege. The UPOV Act 1978 provides Breeders' exemptions. However, the Act 1991 has curtailed Breeders' Exemptions.

(ii) **Farmers' Exemptions:** The legal rights that are provided to farmers to save, use, exchange, share or sell his farm produce of a protected variety are known as farmers' exemptions. Farmers' exemptions are also called Farmers' Rights or Farmers privilege. Here the sale is restricted to non-commercial sale. The UPOV Act 1978 provided farmers Exemptions, however, these exemptions were curtailed by UPOV Act 1991.Because of these reasons, UPOV Act is not accepted by many countries.

Advantages of PBR

1. Breeders get benefit of their variety.
2. PBR help in faster development of seed industry.
3. PBR lead to improvement in quality because of competition.
4. PBR are useful in procurement of good material on payment basis.
5. PBR help in enrichment of genetic resources.

Disadvantages

1. It will promote monopoly.
2. It will encourage unhealthy practices.
3. It may lead to increase in prices.
4. There will be reduction in genetic variability.
5. There will be compulsion to purchase fresh seed every year.

Summary

The Plant Breeders' Rights programme was first established in 1987 under the Plant Variety Rights Act [PVRA], which was succeeded by the current Plant Breeders' Rights Act 1994. {PBRA].Plant breeders' rights [PBR], also known as plant variety rights (PVR), are intellectual property rights granted to the breeder of a new variety of plant. Plant Breeders Rights are granted to novel plant varieties that are distinctive, uniform, and stable (e.g, cultivars breed true-to-type for desired traits). The legal protection of a new plant variety is granted to the breeder or his successor. The effect of PBR is that prior authorization is required before the material can be used for commercial purposes.

Main features of Plant Breeders' Rights in relation to registration, duration, validity, matters covered, requirements, transfer, control, enforcement, etc. have been described.

There are four basic requirements for protection of a variety under Plant Breeders' Rights. These are (i) novelty, (ii) distinctiveness, (iii) uniformity, and (iv) stability

The plant breeders' rights Act provides (i) right to commercial seed production, (ii) right to marketing, (iii) right export and import (iv) right of authorization, and (v) right to prevent infringement for his variety.

The Plant Breeders' Rights Act provides two exemptions, *viz.* (i) breeders' exemptions and (ii) farmers' exemptions. The advantages and disadvantages of Plant Breeders' Rights Act have been outlined.

Questions

1. **Define Plant Breeders' Rights and describe their main features.**
2. **What are advantages of Plant Breeders' Rights?**
3. **Describe briefly disadvantages of Plant Breeders' Rights.**
4. **Explain briefly the following:**
   (i) Breeders Exemption **(ii)** Farmers Exemption
5. **Write short notes on the following:**
   (i) Novelty **(ii)** Distinctiveness
   (iii) Uniformity **(iv)** Stability
6. **Discuss briefly the basic requirements for protection of plant variety under PBR.**
7. **Describe Plant Breeders' Rights on the basis of following:**
   (i) Registration **(ii)** Period of protection
   (iii) Rights covered **(iv)** Material covered

8. Give the full forms of the following:

(i) PVR
(ii) PBR
(iii) IPR
(iv) UPOV

9. Define Infringement. What are the penalties for infringement of PBR ?

10. Discuss briefly the rights granted to a breeder for his variety under PBR.

11. What do the following terms signify?

(i) Breeders' privilege
(ii) Farmers' privilege
(iii) Novelty
(iv) Stability

Chapter 12

Farmers' Rights

Introduction

Farmers' rights refer to the rights arising from the past, present and future contributions of farmers in conserving, improving and making available plant or animal genetic resources, particularly those in the centres of origin/diversity. In other words, Farmers Rights refer to the legal rights provided to farmers to save, use, sow, replant, exchange, share or sell his farm produce including seed of a variety protected under Plant Variety Protection Act. The purpose of these rights is to "ensure full benefits to farmers and support the continuation of their contributions

The FAO Conference held in Rome from 11-29 November, 1989 endorsed the concept of Farmers Rights with a view to:

(i) Ensuring global recognition of the need for conservation and the availability of sufficient funds for these purposes;

(ii) Assisting farmers and farming communities throughout the world, especially those in areas of original diversity of plant genetic resources, in the protection and conservation of their PGR and of the natural biosphere; and,

(iii) Allowing the full participation of farmers, their communities and countries in the benefits derived, at present and in the future, from the improved use of PGR.

Main Features

Main features of Farmers Rights in relation to registration, requirements, fee, rights covered, rights of communities, authorization, materials covered, duration, validity, benefit sharing, gene fund, rewards, infringement etc are briefly presented below;

1. **Registration:** A farmer who has bred or developed a new variety shall be entitled for registration and protection of his variety like a breeder. The registration process consists of the following steps:
   - **(i) Filing of application:** The application is to be filed in the prescribed form and submitted in the office of the registrar of Plant Variety Protection and farmers' Rights Act. The application should contain all the desired documents and information.
   - **(ii) Examination of the application:** The application is examined by experts in the registrar office. If found correct, it is advertised for opposition. Three months time is given for filing opposition.
   - **(iii) Registration:** If there is no opposition, and the registrar and Authority are satisfied, the variety is registered and the registration certificate is issued to the breeder, farmer or other owner of the variety.
2. **Requirements:** The farmers; variety shall be entitled for registration if it meets the following requirements.
   - **(i)** It should have a denomination [name] assigned by the applicant
   - **(ii)** An affidavit that the variety does not contain terminator gene.
   - **(iii)** A complete passport data of parental lines from which the variety has been derived.
   - **(iv)** The contribution of any farmer, village community, institution or organization in the development of the variety.
   - **(v)** The source of genetic material used in developing the variety and the parental material used in developing new variety was lawfully acquired.
   - **(vi)** A brief description of the variety about its novelty, distinctiveness, uniformity and stability.
   - **(vii)** A proof of the right to make the application.
3. **Exemption from Fee:** A farmer or group of farmers or village community shall not be liable to pay any fee in any proceeding before the Authority or Registrar or the Tribunal or the High Court under Plant Variety Protection and Farmers' Rights Act or the rules made there under.
4. **Rights Covered:** The Farmers' Rights Act provide following rights to the farmers in India even for the variety protected under PVP and FR Act.
   - **(i)** Rights to save, use, sow or replant seeds of his farm produce.
   - **(ii)** Rights to exchange or share his farm produce, and
   - **(iii)** Rights to sell seeds of his farm produce

   However, the farmer cannot sell the seeds or planting material under a brand name.
5. **Rights of Communities:** In India, any of the following can file application for registration on behalf of any village or community.

(i) Any person or group of persons [whether actively engaged in farming or not], and

(ii) Any governmental or non-governmental organization.

After completing all formalities, if the Authority is satisfied, the variety can be registered. If due credit has not been given to the farmers or communities who have helped in the development of the variety, the breeder has to pay compensation to the concerned party. Or part of compensation will be deposited in the gene fund.

6. **Authorization of Farmers' Variety:** The authorization for an essentially derived variety from farmers' variety cannot be given by the breeder of such farmers; variety except with the consent of the farmers or group of farmers or community of farmers who have made contribution in the preservation or development of such variety.

7. **Materials Covered:** Farmers' Rights include new varieties of all self pollinated, cross pollinated and asexually propagated varieties.

8 **Duration:** The protection is provided for a definite period. The duration of protection for different plant species is as follows:

   (i) **For Trees and Vines:** Initially for nine years and maximum for 18 years.

   (ii) **For extant varieties:** Initially six years and maximum for 15 years fro the date registration of the variety.

   (iii) **In other cases:** Initially six years and maximum for 15 years fro the date registration of the variety.

9. **Validity:** The protection is valid only in the country where the variety has been registered.

10. **Benefit Sharing:** If the variety has been developed by a particular farmer, he can get benefits of his variety. In case of community rights, the sharing of benefits becomes difficult and the money is deposited in the gene fund.

11. **Gene Fund:** The central government may constitute a Fund to be called National Gene Fund. The money earned from following will be deposited in this fund.

   (i) The benefit sharing received from the breeder of a variety.

   (ii) The annual payable to the Authority in the form of royalty.

   (iii) The amount of compensation received by the Authority.

   Contribution received from amy National or International Organization and other sources.

   The gene fund may be utilized for benefit sharing, rewards and other expenditure related to Plant Variety Protection and Farmers' Rights Act.

12. **Rewards:** A farmer who is engaged in the conservation of genetic resources of land races and wild relatives of economic plants and their improvement through selection shall be entitled for recognition and reward. The expenses toward rewards will be met out from the National Gene Fund.

13. **Acknowledgement by Breeders:** It is compulsory for the breeder of a variety to reveal the origin of the parental lines plant genetic resources used in the development of a variety. It will help in deciding the share of different persons, *viz.*, farmers and other communities in benefits earned by breeder of such variety.
14. **Infringement:** If a farmer was not aware of the PVP and FR Act at the time of using any protected variety, he will be exempt from the infringement and will be asked not to do such act in future. This right has been provided to Indian farmers by PVP and FR Act 2001.

Nature and Scope of Farmers' Rights

Farmers' Rights, as endorsed by FAO in 1989, recognize that farmers and rural communities have contributed and continue to contribute to creation, conservation, exchange and enhancement of genetic resources and they should be recognized and strengthened in their work. The important points about nature and scope of Farmers' Rights are briefly presented as follows:

1. **Ancient Rights:** Farmers' Rights have a deep historic character. They existed since humans created agriculture to serve their needs, and played vital role in conservation of biodiversity. Farmers are the guardians of these genetic resources, which support the evolution of species. Farmers are the inheritors of the skills and knowledge of the generations that have created this biological wealth. For these contributions, it is necessary to recognize Farmers' Rights.
2. **Rights Covered:** Farmers' Rights include the right over resources and associated knowledge, united indivisibly, and mean the acceptance of traditional knowledge, respect for cultures and recognition that these are the basis of the creation of knowledge.
3. **Property Rights:** The farmers deserve right to control, the right to decide the future of genetic resources, the rights to define the legal framework of property rights of these resources.
4. **Collective Nature:** Farmers' Rights are of an eminently collective nature and therefore are protected by sui generic rights.
5. **Application:** These rights should have a national application, respecting the sovereignty of each country, to establish local laws based on these principles.
6. **Rights to Means:** It includes territorial rights, right to land, right to water and air and rights to conserve biodiversity and achieve food security.
7. **Rights for Active Participation:** The farmers deserve right to participation in the definition, elaboration and execution of policies and programmes linked to genetic resources.
8. **Rights to Technology:** The farmers deserve right to appropriate technology as well as participation in the design and management of research programmes.

9. **Rights to Benefit Sharing:** The farmers deserve right to define the control and handling of benefits derived from the use, conservation and management of these resources.
10. **Rights on Genetic Resources:** The farmers deserve right to use, choose, store and freely exchange of genetic resources.
11. **Rights for Sustainable Agriculture:** The farmers deserveright to develop models of sustainable agriculture that protects the biodiversity and to influence the policies that support it.

Role of Farmers Related to PGR

Farmers are both men and women who have domesticated, developed, conserved and made available plant genetic and other natural resources (land; water; vegetation; animal, bird and fish life) to which they have access in order to obtain a livelihood and ensure the well-being of their family through the provision of basic requirements such as food, fuel and water and income from the land.

Farmers' Rights are not assigned to specific varieties, types of plants, or to specific farmers. Their purpose is to encourage farmers and farming communities to nurture and conserve, and to utilize and improve, plant genetic resources (FAO, 1996).

Farmers' Rights are generally considered to be group (collective) rights that do not necessarily imply direct compensation or bilateral reciprocity, such as selling a license for exclusive rights to collect and use crop germplasm.

Farmers have played key role in conservation of Plant Genetic Resource for thousands of years. The role of farmers in relation to PGR is briefly presented below:

1. Farmers both men and women have conserved plant genetic resources over the past thousands of years.
2. Farmers are rightly called as on site managers of plant genetic resources.
3. Farmers have in depth knowledge of the land races and wild relatives of cultivated plant species.
4. All modern plant varieties contain only those genes that have originated from farmers' traditional varieties (landraces) or wild crop relatives
5. The women and men have historically provided the germplasm upon which scientific plant breeding is based.
6. Generally, germplasm is provided by small-scale subsistence farmers to plant breeders whose varieties are then adapted/developed for medium to large scale commercial farmers (Spillane, 1996).
7. Women's special knowledge of the value and diverse uses of plants for nutrition, health and income has important implications for the conservation of plant genetic resources.
8. The women have played significant role not only in terms of labour and skills, but also their decision-making about how natural resources are used to satisfy the multiple needs of rural households.

9. In much of the developing world, the on-farm and in-situ conservation and use of plant genetic resources begins with women. As farmers, they are responsible for growing crop plants and collecting seed for next year sowing.
10. It is estimated that ninety per cent of the planting materials used in developing countries is produced by the farmers themselves, *i.e.* from saved seed (Spillane, 1996). The selection of wild genetic resources for home planting, of seeds for conserving for next year's planting and of asexually propagated material is usually carried out by women. This selection is a sophisticated process that takes into account many different criteria such as taste, color, palatability and texture; resistance to pests and diseases; adaptation to soil and agro-climatic conditions and so on. The selection criteria are built up over years of experience.

Comparison of PBR and Farmers Rights

Plant Breeders' Rights and Farmers' Rights have some similarities and some difference. A brief comparison of these two is presented in Table 12.1.

Table 12.1: Comparison of Plant Breeders' Rights and Farmers Rights

Sl.No.	*Particulars*	*Plant Breeders' Rights*	*Farmers' Rights*
A.	**Similarities**		
1.	Registration	Essential	Essential
2.	Duration		
	Trees and vines	18 years	18 years
	Extant varieties etc.	15 years	15 years
3.	Enforcement	After Registration	After Registration
4.	Requirements	Novelty, Distinctiveness, uniformity and stability	Same as in PBR
5.	Validity	In country of registration	In country of registration
B.	**Dissimilarities**		
1.	Registration Fee	Charged	Not charged
2.	Annual renewal fee	Charged	Not charged
3.	Penalty for infringement	Charged	Not charged
4.	Community Rights	Not covered	Covered
5.	Authorization	Easy	Not permitted

Advantages of Farmers' Rights

1. Provide recognition to farmers, communities or village for their role in conserving Plant Genetic Resources.
2. Provide respect to farmers.
3. Provide opportunities to farmers to participate in the policies, projects and programs related to plant Genetic Resources [PGR].

4. Lead to benefit sharing when farmers material is used in developing new variety by breeders.
5. Farmers permission is required to use their plant material.
6. Provide rewards to those farmers who are actively engaged in conserving Plant Genetic Resources.

Limitations of Farmers' Rights

1. Mostly farmers' rights are of collective nature.
2. The benefit sharing is difficult.
3. Farmers do not get any financial support from government for the conservation of Plant Genetic Resources.
4. Farmers' Rights are of territorial nature.

Summary

The concept of Farmers' Rights was endorsed by the FAO Conference held in Rome from 11-29 November, 1989. Farmers' rights refer to the rights arising from the past, present and future contributions of farmers in conserving, improving and making available plant or animal genetic resources, particularly those in the centres of origin/ diversity. In other words, the legal rights provided to farmers to save, use, sow, resow, exchange, share or sell his farm produce including seed of a variety protected under Plant Variety Protection Act refer to Farmers Rights.

The purpose of these rights is to "ensure full benefits to farmers and support the continuation of their contributions.

Main features of Farmers Rights in relation to registration, requirements, fee, rights covered, rights of communities, authorization, materials covered, duration, validity, benefit sharing, gene fund, rewards, infringement etc have been discussed.

The important points about nature and scope of Farmers' Rights have been explained. The role of farmers in relation to conservation of plant genetic resources has been amply discussed. The comparison of Plant Breeders' Rights and Farmers' Rights has been given in tabular form. The advantages and limitations of Farmers' Rights are listed and future outlook of plant genetic resources have been discussed.

Questions

1. **Define Farmers' Rights and describe their significance.**
2. **Describe briefly main features of Farmers' Rights.**
3. **Explain briefly the procedure of registration of Farmers' variety**
4. **Describe Farmers' Rights on the basis of following:**

   (i) Registration
   (ii) Duration of protection
   (iii) Rights covered
   (iv) Authorization

5. **Write short notes on the following:**
   - **(i)** *In situ* conservation
   - **(ii)** *Ex situ* conservation
   - **(iii)** Biodiversity
   - **(iv)** Land Race
6. **What do the following terms signify**
   - **(i)** Farmers' Rights
   - **(ii)** Genetic Resources
   - **(iii)** Public Domain
   - **(iv)** Cultivar
7. **Give a brief comparison of Plant Breeders' Rights and farmers' Rights in a tabular form.**
8. **Describe the role of farmers in relation to Plant Genetic resources.**
9. **Discuss the following in relation to Farmers' Rights.**
   - **(i)** Infringement
   - **(ii)** Registration Fee
   - **(iii)** Role of women farmers
   - **(iv)** Benefit sharing
10. **Describe the role of women farmers in relation to Plant Genetic resources.**
11. **Describe briefly ways and means for sustainable use of Plant Genetic Resources.**

Section III

Biodiversity Related Legislations

Chapter 13

Convention on Biodiversity

Introduction

The biodiversity refers to the variety of life in all forms, levels and combinations. The term biodiversity includes genetic diversity, species diversity, and ecosystem diversity. Crop biodiversity refers to the variety of genes and genotypes found in crop plants. Biodiversity is a modern term which simply means "the variety of life on earth". This variety of life can be measured in several ways.

1. Genetic Diversity

Genetic diversity refers to the variation between individuals of the same species. This includes genetic variation between individuals in a single population, as well as variations between different populations of the same species. The main points related to genetic diversity are listed below:

(i) It represents variation between individuals of the same species.
(ii) It represents the variation at gene level.
(iii) It can be measured at population, species and community levels.
(iv) Sophisticated techniques are required for its measurement.
(v) It represents the raw material for evolution and adaptation.
(vi) It is effected by the environment and competition with other species.

2. Species Diversity

Species diversity refers to the variety of species in a given region or area. This can either be determined by counting the number of different species present, or by determining taxonomic diversity. Taxonomic diversity is more precise and considers the relationship of species to each other. For example, an area containing three species of wheat and two species of barley, is more diverse than an area containing five species of wheat, even though they both contain the same number of species. The main points related to species diversity are listed below:

(i) It represents variety of species in a given region or area.

(ii) It identification is easy by naked eye in the field.

(iii) It can be determined simple by counting the number of different species.

(iv) Species are distinct units of diversity.

3. Ecosystem Diversity

Ecosystem diversity refers to communities of plants and animals, together with the physical characteristics of their environment (*e.g.* geology, soil and climate). Ecosystem diversity is more difficult to measure because there are rarely clear cut boundaries between different ecosystems and they grade into one another. However, if consistent criteria are chosen to define the limits of an ecosystem, then their number and distribution can also be measured. The main points related to ecosystem biodiversity are given below.

(i) It represents communities of plants and animals together with physical structures of their environment.

(ii) The measurement of ecosystem diversity is more difficult than genetic diversity and species diversity.

(iii) The assessment of ecosystem diversity needs many complex measurements to be taken over a long period of time

(iv) It is a time consuming task and completion of such work requires adequate staff and financial support.

Advantages of Crop Biodiversity

There are several advantages of crop biodiversity. Some important advantages or benefits of crop biodiversity are briefly presented as follows. The crop biodiversity provides Choice of items for food, fibre, oil, fuel, fodder to animals, timber and medicines. It provides protection from biotic [disease and insects]and abiotic [drought, soil salinity, soil alkalinity, soil acidity, cold temperature, heat etc.] stresses and also provides wider adaption and broad genetic base to the populations.

Losses of Crop Biodiversity

There are two main reasons of depletion of biodiversity in crop plants, *viz.*, extinction and genetic erosion. Extinction refers to permanent loss of a crop species due to various reasons. Genetic erosion refers to the gradual reduction in genetic variability, in population of a species, due to elimination of various genotypes. There are five main reasons of genetic erosion, *viz.* (i) replacement of land races by modern cultivars, (ii) industrial agriculture and mechanization, (iii) farming into wild habitat, (iv) clean cultivation and (v) developmental activities.

Conservation of Biodiversity

Conservation refers to protection of genetic diversity of crop plants from genetic erosion. There are two major methods of conserving *viz*. (i) *in-situ* conservation and (ii) Ex-situ conservation.

(i) *In situ* Conservation

In situ (=on-site) conservation and use refers to the maintenance and use of wild plant populations in the habitats where they naturally occur and have evolved without the help of human beings. In other words, it refers to conservation of biodiversity under natural habitats. In this method, the wild species and the complete ecosystems are preserved together.

The main focus of Biodiversity's work is on the maintenance of the genetic diversity of the wild species and wild relatives of crop plants in the protected areas [national parks and nature reserves]. The advantages and disadvantages of this method are presented below:

Advantages

The biodiversity is conserved under natural habitats.It improves of the populations. It conserve complete ecosystem means several species together.It protects endangered species in their original habitats.

Disadvantages

It requires establishment of gene sanctuaries and national parks.Several areas have to be conserved to protect a single species.The management of protected ares poses several problems. It is costly method of biodiversity conservation.

(ii) *Ex situ* Conservation

Ex situ (= off-site) conservation of germplasm takes place outside the natural habitat or outside the production system, in facilities specifically created for this purpose. Depending on the type of species to be conserved, different *ex situ* conservation methods may be used. The gene banks are used for ex situ conservation of biodiversity.In gene banks, biodiversity is conserved for its sustainable use by breeders, farmers and researchers for agricultural development. The advantages and disadvantages of this method are given below:

Advantages

One species can be conserved at one place.The biodiversity is conserved either in seed gene banks or in field gene banks. Handling of material is easy. It requires less space for conservation.

Disadvantages

It requires establishment of seed gene banks and field gene banks.It require regular funding for maintenance of gene banks. It involves high technical man power. It requires periodical rejuvenation in case of seed crops.

Convention on Biological Diversity [CBD]

The Convention on Biological Diversity is also known as the Biodiversity Convention. It is an international treaty that was adopted in Rio de Janeiro in June 1992. The Convention was opened for signature at the Earth Summit in Rio de Janeiro on 5 June 1992 and entered into force on 29 December 1993. The biodiversity agreement has been signed by 189 countries. The Convention has following three main goals:

1. Conservation of biological diversity (or biodiversity);
2. Sustainable use of its components; and
3. Fair and equitable sharing of benefits arising from genetic resources.

1. Conservation of Biodiversity

The convention recognized for the first time in international law that the conservation of biological diversity is "a common concern of humankind" and is an integral part of the development process. Its objective is to develop national strategies for the conservation and sustainable use of biological diversity. The Convention recognizes that ecosystems, species and genes must be used for the benefit of humans. However, this should be done in a way and at a rate that does not lead to the long-term decline of biological diversity.

The convention states that if there is a threat of significant reduction or loss of biological diversity, immediate measures must be taken up to avoid or minimize such a threat. The conservation of biodiversity, though requires substantial investments, will bring us significant environmental, economic and social benefits in return.

2. Sustainable Use

The agreement covers all ecosystems, species, and genetic resources. It links traditional conservation efforts to the economic goal of using biological resources sustainably. The convention reminds decision-makers that natural resources are not infinite and sets out a philosophy of sustainable use.

3. Equitable Benefit Sharing

It sets principles for the fair and equitable sharing of the benefits arising from the use of genetic resources, notably those destined for commercial use. The main points are as follows:

1. Measures and incentives for the conservation and sustainable use of biological diversity.
2. Regulated access to genetic resources and traditional knowledge, including Prior Informed Consent of the party providing resources.
3. Sharing, in a fair and equitable way, the results of research and development and the benefits arising from the commercial and other utilization of genetic resources.
4. Access to and transfer of technology, including biotechnology, to the governments and/or local communities that provided traditional knowledge and/or biodiversity resources.

5. Technical and scientific cooperation.
6. Impact assessment.
7. Education and public awareness.
8. Provision of financial resources.
9. National reporting on efforts to implement treaty commitments.

Summary

The biodiversity refers to the variety of life in all forms, levels and combinations. The term biodiversity includes genetic diversity, species diversity, and ecosystem diversity. It is usually considered at three levels, *viz.*, genetic diversity, species diversity and ecosystem diversity. Crop biodiversity refers to the variety of genes and genotypes found in crop plants. There are three levels of biodiversity, *viz.*, genetic diversity, species diversity, and ecological diversity. These have been discussed.

There are different approaches to conserving biodiversity and different ways of using genetic resources. These are (*i*) on farm management, (*ii*) *In situ* conservation, (*iii*) *Ex situ* conservation and (*iv*) complementary conservation.

The Convention on Biological Diversity is also known as the Biodiversity Convention. It is an international treaty that was adopted in Rio de Janeiro in June 1992. The Convention was opened for signature at the Earth Summit in Rio de Janeiro on 5 June 1992 and entered into force on 29 December 1993. The biodiversity agreement has been signed by 189 countries. The Convention has three main goals, *viz.* (i) conservation of biological diversity (or biodiversity), (ii) sustainable use of its components; and (iii) fair and equitable sharing of benefits arising from genetic resources. The secretariat of CBD is in Montreal, Canada.

Questions

1. **Define biodiversity and describe briefly its advantages.**
2. **What is plant biodiversity ? Give a brief account of different type of crop plants diversity.**
3. **Define genetic diversity and describe briefly it significance.**
4. **Describe briefly various reasons of loss in biodiversity.**
5. **What do the following terms signify?**

   (i) Genetic diversity | **(ii)** Species diversity
   (iii) Ecosystem diversity | **(iv)** Land Races
   (v) Adaptation | **(vi)** Extinction
   (vii) Seed Bank | **(viii)** Gene Bank

6. **Write short notes on the following:**

   (i) *In situ* conservation
   (ii) *Ex situ* conservation
   (iii) Orthodox seeds
   (iv) Recalcitrant seeds

7. **Describe briefly various ways of conserving crop biodiversity.**

8. **Differentiate between the following:**

   (i) Genetic diversity and species diversity
   (ii) Land Races and Modern cultivars
   (iii) Orthodox seeds and recalcitrant seeds
   (iv) *In situ* conservation and Ex situ conservation

4. **Describe the role of CBD in relation to conservation of biodiversity.**

Chapter 14

Indian Biodiversity Legislation

Introduction

In India, the Biological Diversity Act [also called biodiversity act] was passed by the Central Government in 2002.However, it came into force on 29th December, 2003. Indian biodiversity Act, 2002 consists of 12 chapters. Contents of all these 12 chapters are briefly presented in the tabular form as follows:

Table 14.1: Brief Description of the Chapters of Indian Biodiversity Act 2002

Sl.No.	*Chapter*	*Brief Explanations*
1.	Chapter 1	It explains that the act will be called Biological Diversity Act, 2002. It extends to whole of India. Various terms such as benefit claimers, biological diversity, biological resources, bio-survey, chairperson, commercial utilization, fair and equitable benefit sharing, local bodies, member, national biodiversity authority, regulations, research state biodiversity board, sustainable use and value added products have been defines.
2.	Chapter 2	It deals with regulation of biological diversity. It explains items which require approval of National Biodiversity Authority and which do not require permission. It also explains people who require permission and who do not require permission.
3.	Chapter 3	It deals with establishment of National Biodiversity Authority, its headquarters, constitution of National Biodiversity Authority, its service conditions (tenure), its sub-committees, delegation of powers, etc.
4.	Chapter 4	It deals with functions and powers of National Biodiversity Authority such as issue of guide line for equitable benefit sharing. Authority may advise to the Central/ State Government on matters related to biodiversity. The authority may take any measure to oppose grant of IPR to any other country on behalf of Government of India.

Contd...

Table 1.1–*Contd...*

Sl.No.	*Chapter*	*Brief Explanations*
5.	Chapter 5	This chapter deals with approval of National Biodiversity Authority for undertaking certain activities, transfer of biological resource or knowledge and determination of equitable benefit sharing.
6.	Chapter 6	This chapter deals with constitution of State Biodiversity Board. It also defines its functions and powers.
7.	Chapter 7	This chapter deals with finance, accounts and audit of National Biodiversity Authority. It also deals with preparation of annual report of National Biodiversity Authority. The report is presented before Parliament.
8.	Chapter 8	This chapter deals with finance, accounts and audit of State Biodiversity Board. It deals with preparation of annual report of State Biodiversity Authority. The report is presented before the House of State Legislature.
9.	Chapter 9	This chapter deals with duties of Central and State governments. Central government has to develop national strategies, plans, etc. for conservation of biological diversity. It also decides heritage sites, identify institutions as repositories. It also explains items that are exempted from biodiversity act.
10.	Chapter 10	This chapter deals with constitution of Biodiversity Management Committees.
11.	Chapter 11	This chapter deals with constitution of local biodiversity fund and its applications. It also deals with audit of accounts of biodiversity Management Committees, preparation of annual reports of such committees and submission of report to the District Magistrate.
12.	Chapter 12	This chapter deals with power of State government to give directions, settlement of disputes between State Biodiversity Boards, appeals, penalties, offences by companies, power of central government to give directions to State Government, Power of central government to make rules, power of State government to make rules, power to make regulations, power to remove difficulties, etc.

Salient Features

The salient features of the Biodiversity Legislation such as purpose, structure, research on biological resources, Indian industry and biodiversity act, Ayurvedic industry, biodiversity act versus PPV and FR act, heritage sites, threatened species, repositories, check on bio-piracy and benefit sharing are briefly presented below:

1. **Purpose:** The purpose of The Biodiversity Act - 2002 is to protect India's rich biodiversity [genetic resources] and associated knowledge against their use by foreign individuals, institutions or companies without sharing the benefits arising out of such use, and check bio-piracy.
2. **Structures of Biodiversity Act – 2002:** The biodiversity Act provides for establishment of a three tiered structure at the national, state and local level. The three components of Indian Biodiversity Act are (i) National Biodiversity Authority (NBA), (ii) State Biodiversity Boards (SBB) and (iii) Biodiversity Management Committees (BMCs. These are briefly discussed as follows.

   (i) National Biodiversity Authority (NBA): All matters relating to requests for access by foreign individuals, institutions or companies, and all matters relating to transfer of results of research to any foreigner will be dealt with by the National Biodiversity Authority.

(ii) **State Biodiversity Boards (SBB)**: All matters relating to access by Indians for commercial purposes will be under the purview of the State Biodiversity Boards (SBB). The Indian industry will be required to provide prior intimation to the concerned SBB about the use of biological resource. The State Board will have the power to restrict any such activity, which violates the objectives of conservation, sustainable use and equitable sharing of benefits.

(iii) **Biodiversity Management Committees (BMCs)**: Institutions of local self government will be required to set up Biodiversity Management Committees in their respective areas for conservation, sustainable use, documentation of biodiversity and chronicling of knowledge relating to biodiversity. NBA and SBBs are required to consult the concerned BMCs on matters related to use of biological resources and associated knowledge within their jurisdiction.

3. **Research on Biological Resources:** There is no requirement under the legislation for seeking permission of the National Biodiversity Authority for carrying out research, if it is carried out in India by Indians, as well as under collaborative research projects that have been drawn within the overall policy guidelines formulated by the Central Government. The only situations that would require permission of the NBA are: (i) when the results of any research which has made use of the country's biodiversity is sought to be commercialized, (ii) when the results of research are shared with a foreigner or foreign institution, and (iii) when a foreign institution/individual wants access to the country's biodiversity for undertaking research.

4. **Exemptions:** The legislation provides for the following exemptions.

(i) Exemption to local people and community of the area for free access to use biological resources within India.

(ii) Exemptions to growers and cultivators of biodiversity and to Vaids and Hakims to use biological resources.

(iii) Exemption through notification of normally traded commodities from the purview of the Act.

(iv) Exemption for collaborative research through government sponsored or government approved institutions subject to overall policy guidelines and approval of the Central Government.

(v) The value added products have been excluded from the definition of biological resources.

(vi) The Indian researchers neither require prior approval nor need to give prior intimation to SBB for obtaining biological resource for conducting research in India.

5. **Indian industry and Biodiversity Act:** The Indian industry is required to give prior intimation to the concerned SBB about obtaining the biological resources for commercial purposes. The SBB will have the power to prohibit or restrict any such activity, which violates the objectives of conservation, sustainable use and equitable sharing of benefits.
6. **Ayurvedic Industry:** The Act does not aim at banning the use of medicinal plants. It provides that for commercial use of resources and related knowledge by Indians only, prior intimation to the State Biodiversity Board is required. Hakims and vaids will continue to have free access to resources and knowledge.
7. **Biodiversity Bill and Plant Varieties Protection Act:** There is no overlap between Biodiversity Bill and Plant Varieties Protection (PVP) Act The scope and objectives of these two legislations are different The PVP legislation accords intellectual property rights to a person for developing a new plant variety. On the other hand, the biodiversity legislation is primarily aimed at regulating access to biological resources and associated knowledge so as to ensure equitable sharing of benefits arising from their use.
8. **Heritage sites:** The Act [Section 37] provides for designating heritage sites. These are areas of biodiversity importance, which harbour rich biodiversity, wild relatives of crops, or areas, which lie outside the protected area network. The purpose is not to cover the already designated protected areas such as national parks and wildlife sanctuaries.
9. **Threatened species:** The Act [Section 38] provides for notifying threatened species and prohibits or regulates their collection. It also provides for taking appropriate steps to rehabilitate and preserve those species, thereby ensuring their conservation and management.
10. **Repositories:** The Act [Section 39] provides for designating some premier institutions of the country (*e.g.* Botanical Survey of India, Zoological Survey of India, National Bureau of Plant Genetic Resources etc.) as repositories for keeping different categories of biological resources for safe custody and posterity.
11. **Check on Bio-piracy:** To check bio piracy, the proposed legislation provides that access to biological resources and associated knowledge is subject to terms and conditions, which secure equitable sharing of benefits. Further, it would be required to obtain the approval of the National Biodiversity Authority before seeking any IPR based on biological material and associated knowledge obtained from India.
12. **Benefit Claimers:** The benefit claimers are conservers of biological resources, creators and holders of knowledge and information relating to the uses of biological resources. While granting approvals, NBA will impose conditions, which secure equitable share in the benefits arising out of the use of biological resources occurring in India or knowledge relating to them. These benefits could include monetary gains, grant of joint ownership of IPRs, transfer of technology, association of Indian Scientists in R and D, setting up of venture capital fund etc.

In cases where specific individuals, or group of individuals are identifiable, the monetary benefits will be paid directly to them. Otherwise, the amount will be deposited in the National Biodiversity Fund.

Advantages of Biodiversity Act

There are several advantages of Indian biodiversity act 2002. Important advantages are listed as follows:

1. It will help in conservation of plant and animal diversity available in the country for the use of present and future generations.
2. It will help in equitable benefit sharing arising from the use national biodiversity to those who are responsible for the maintenance of biodiversity directly or indirectly.
3. It will also help in sharing benefit arising from the use of knowledge associated with biological resources of the country.
4. It will protect the biological diversity from the exploitation of foreign countries. Companies of foreign countries will have to pay for making use of Indian biodiversity.
5. It has provision for proper maintenance of national biodiversity. The biodiversity will be maintained in the government institutions identified for such purpose.
6. It permits use of biodiversity for preparation of Ayurvedic medicines and also for research purposes by research institutions. There is no need of seeking permission of Biodiversity Authority for carrying research in India or in collaborative international research projects formulated by central government.
7. It will help in sustainable use of biodiversity. It does not permitted exploitation of resources. Biodiversity has to be used in such a way that it should be available for the present generation as well as for the future generations.

Summary

In India the Biological Diversity Act [also called biodiversity act] was passed by the Central Government in 2002.The salient features of the Biodiversity Legislation such as purpose, structure, research on biological resources, Indian industry and biodiversity act, Ayurvedic industry, biodiversity act versus PPV and FR act, heritage sites, threatened species, repositories, check on bio-piracy and benefit sharing have been discussed.

The biodiversity Act provides for establishment of a three tiered structure at the national, state and local level. The three components of Indian Biodiversity Act are (i) National Biodiversity Authority (NBA), (ii) State Biodiversity Boards (SBB) and (iii) Biodiversity Management Committees (BMCs. These are briefly discussed. Advantages of Indian biodiversity act 2002 have been listed.

Questions

1. **Describe briefly the purpose and structure of Indian biodiversity act 2002.**

2. **Explain briefly exemptions of Indian Biodiversity Act 2002.**

3. **Discuss Indian Biodiversity Act 2002 in relation to Ayurvedic industry and benefit sharing.**

4. **Explain briefly Indian Biodiversity Act 2002 in relation to heritage sites, repositories and check on bio-piracy.**

5. **Write short notes on the following:**

   (i) Heritage sites (ii) Repositories

   (iii) Check on bio-piracy (iv) Exemptions of biodiversity act

4. **Explain the role of following in relation to Indian Biodiversity Act 2002.**

   (i) National Biodiversity Authority (NBA)

   (ii) State Biodiversity Boards (SBB)

   (iii) Biodiversity Management Committees (BMCs)

5. **Describe briefly salient features of Indian Biodiversity Act, 2002.**

6. **Explain briefly Indian Biodiversity Act 2002 in relation to Protection of Plant Varieties and Farmers' Rights Act 2001.**

7. **Describe the role of CBD in relation to conservation of biodiversity.**

8. **Describe advantages of the Indian Biodiversity Act.**

Section IV

Bio-safety Related Legislations

Chapter 15

Cartagena Protocol on Bio-safety

Introduction

The Cartagena Protocol on Bio-safety is the first international agreement to regulate the trans-boundary movements of genetically engineered (GE) organisms. The Bio-safety Protocol is a subsidiary agreement to the UN Convention on Biological Diversity (CBD), which was signed by over 150 governments at the Rio Earth Summit in 1992. The Bio-safety Protocol seeks to protect biological diversity from the potential risks posed by living modified organisms [LMOs] resulting from modern biotechnology.

Bio-safety refers to the safe application of modern biotechnology with regard to human health, animal health and environment. In other words, bio-safety is a term used to describe efforts to reduce and eliminate potential risk resulting from biotechnology and its products.

Brief History

The Cartagena Protocol arises from the United Nations Conference on Environment and Development (also known as the Earth Summit) held in Rio de Janeiro, Brazil, in June 1992, in which Convention on biological diversity was signed. The Cartagena Protocol on Bio-safety, also known as the Bio-safety Protocol, is a supplement of the convention on biological diversity. The Bio-safety Working Group met six times between 1996 and 1999. The extraordinary meeting of the conference of parties [COP], was originally scheduled to adopt the Protocol in February 1999 in Cartagena city of Colombia. However, the meeting broke down because of disagreements regarding appropriate consideration of economic interests. The extraordinary meeting was temporarily suspended and only resumed after three more negotiation rounds achieved compromise. The Protocol was finally adopted by conference of parties [COP] in Montreal, Canada on 29 January 2000 and entered into force on 11 September 2003.

Need for Bio-safety Protocol

The need of such a strong Bio-safety Protocol is illustrated by the genetic contamination of maize in Mexico. Genetically engineered cultivars of maize in Mexico resulted in contamination of non-transgenic varieties, land races and wild relatives of this major food crop. This was the first case of genetic pollution in a centre of origin and diversity of a major food crop. This invited the attention of crop scientists for adoption of strong bio-safety protocol for handling of LMOs.

Member Countries

By June 2001, the Protocol had received 103 signatures. The membership is open to all countries of the World. The membership has been signed by 162 countries till November 2011. India signed the Cartagena Protocol in January 2001. However, countries like US, Argentina and Canada that produce about 90 percent of genetically engineered [GE] crops in the world have not ratified the Protocol, and are actively working to undermine it.

Entry in to Force

The required number of 50 instruments of ratification/accession/approval/ acceptance by countries was reached in May 2003. As a result, the Protocol entered into force on 11 September 2003.

Governing Body

The governing body of the Protocol is the Conference of the Parties to the Convention serving as the meeting of the Parties to the Protocol (COP-MOP). The main function of this body is to review the implementation of the Protocol and make decisions necessary to promote its effective operation. As of date, four meetings of the COP-MOP have been convened. The governing body includes Chairman and members from member countries. The Chairman is of the rank of Ambassador or High Commissioner of the member country. The term of the governing body is usually five years. The secretariat of the protocol is the same as for convention on biological diversity [CBD]. It is located at Montreal in Canada. Funds for the protocol are voluntarily contributed by contracting parties, developed country parties and other countries and sources. The main functions of governing body are:

1. To review the implementation of the Protocol and
2. To make decisions necessary to promote its effective operation.

Functions of Bio-safety Protocol

Bio-safety protocol refers to various measures that are taken for the safe application of modern biotechnology. These are regulatory measures and policies that are designed to safeguard the public interest. The bio-safety protocol became effective from January 2000. The main functions of bio-safety protocol are as follows.

1. **Promotion of Bio-safety:** The Protocol promotes bio-safety measures by establishing rules and procedures for the safe transfer, handling, and use of LMOs, developed through modern biotechnology.
2. **Conservation of Biodiversity:** The Protocol applies to the trans-boundary movement [transit, handling and use] of all living modified organisms that may have adverse effects on the conservation and sustainable use of biological diversity. It ensures that LMOs should not have adverse effects on human health, animal health and biological diversity.
3. **Regulation of World Trade:** Protocol is designed to regulate the international trade of genetically engineered organisms that may have adverse effects on the conservation and sustainable use of biological diversity, including risk to human health.

Objectives and Elements of Protocol

The objective of the Protocol is to ensure an adequate level of protection in the field of the safe transfer, handling and use of 'living modified organisms resulting from modern biotechnology' that may have adverse effects on the conservation and sustainable use of biological diversity, including risks to human health, specifically from trans-boundary movement of LMOs. The main elements of Cartagena bio-safety protocol are briefly presented here as under.

1. **Clearing House:** The Protocol establishes an Internet-based "Bio-safety Clearing-House" to help countries exchange scientific, technical, environmental, and legal information about living modified organisms (LMOs).
2. **Advance Informed Agreement Procedure**: It creates an advance informed agreement (AIA) procedure that in effect requires exporters to seek consent from an importing country before the first shipment of an LMO meant to be introduced into the environment (such as seeds for planting, fish for release, or microorganisms for bioremediation).

   The AIA procedure does not apply to LMO commodities that are intended for food, feed, or processing (*e.g.*, maize, soy or cottonseed), to LMOs in transit, or to LMOs destined for contained use (*e.g.*, organisms intended only for scientific research within a laboratory).
3. **Items not Addressed:** The Protocol does not address food safety issues. It does not pertain to non-living products derived from genetically engineered plants or animals, such as milled maize or other processed food products.
4. **Documentation:** It requires shipments of LMO commodities, such as maize or soybeans that are intended for direct use as food, feed, or for processing, to be accompanied by documentation stating that such shipments "may contain" living modified organisms and are "not intended for intentional introduction into the environment. It also sets out information to be included on documentation accompanying LMOs destined for contained use, including any handling requirements and contact points for further information and name and address of the importer and exporter. It should

also include a declaration that the movement is in conformity with the Protocol and, as appropriate.

5. **Existing Rights Remain Unchanged:** Parties must implement rights and obligations under the Protocol consistent with their existing international rights [such as WTO agreements] and obligations, including with respect to non-Parties to the Protocol.
6. **Capacity Building:** The Protocol calls on Parties to cooperate with developing countries in building their capacity for managing modern biotechnology.
7. **Trade with non-Parties:** The Protocol states that the "trans-boundary movement of living modified organisms between Parties and non-Parties shall be consistent with the objective of this Protocol.

Risks from GMOs and their Products

There are two broad categories of risk from genetically modified organisms and their products, *viz.* (i) risk to human and animal health, and (ii) risk to the environment. These aspects are briefly discussed as follows:

1. Risk to Human and Animal Health

The risk from genetically modified organisms and their products to human health include toxicity, allergenicity, antibiotic resistance and nutritional uptake. These are discussed as follows:

(i) **Toxicity:** Sometimes, the transgene leads to production of toxic substance by altering the metabolic pathway. Consumption of such transgenic products may cause toxic effect on human health. Hence, every transgenic product has to be evaluated for toxicity to human and animal health.

(ii) **Allergenicity**: Sometimes, transgenic products have allergic effects on human health. For example, transgenic soybean containing methionine gene from Brazil nut has allergic effects in sensitive persons. Hence, such soybean has not been approved for sale. Feeding of Bt. cotton seed to animals has not been reported to have any adverse effect. The Bt. cake also does not have any adverse effect on the digestion of animals. Moreover, no allergic or toxic effect of Bt. cotton seed and meal has been reported so far. The oil extracted from the seed of Bt. cotton has not been found to have any adverse effect on the human health.

(iii) **Antibiotic Resistance**: Antibiotic such as Kanamycin is used for selection of transgenic cells. It is believed that transgenic products when consumed will lead to resistance to such antibiotic. However, there is no such report so far.

(iv) **Nutrient Uptake**: There are possibilities that transgene product may inhibit absorption of some nutrients in human digestive system, resulting in adverse effects on human health. However, such evidences are not available.

2. Risk to Environment

The risk from genetically modified organisms and their products to environment include development of super weeds, loss of biodiversity, contamination of non-transgenic cultivars, effect on non-target organisms, development of resistance, effect on soil ecology and evolution of new viruses. These aspects are briefly discussed as follows:

(i) **Development of Super Weeds:** Natural out crossing of transgenic plants with wild species and relatives of a crop may lead to development of super weeds. Such weeds may have herbicide resistance or insect resistance and will be more invasive than normal weeds. The out crossing is easy with other varieties or cross compatible species.

(ii) **Loss of Biodiversity**: Some farmers still maintain landraces and some old varieties which are valuable genetic resources. The natural ecosystem is still rich in plant biodiversity which needs to be conserved. The wide spread use of transgenic cultivars of different crops will lead to substantial loss of crop biodiversity.

(iii) **Contamination:** The cultivation of transgenic varieties will lead to contamination of landraces, non-transgenic cultivars and closely related wild species through natural out crossing. In Mexico, great diversity of both wild and cultivated corn is found. It has been reported that wild corn varieties in some remote areas of Mexico has been contaminated by Bt. transgenic cultivars.

The possibilities of cross pollination of Bt. cotton with other species are nil to negligible because the Bt. gene has been inserted in upland cotton ($2n = 52$) which cannot outcross with cultivated or wild diploid species ($2n = 26$). It can also not outcross with tetraploid wild species such as *G. tomentosum* which are found either in uncultivated areas or extremely isolated species garden maintained at different research institutes. The upland cotton in which Bt. gene has been inserted does not have cross compatibility with other genera of the family *Malvaceae.*

(iv) **Effect on Non-target Insects:** The genetically modified genes may sometimes have adverse effects on non-target insects/organisms. The adverse effects of Bt. insecticidal protein have been reported. No adverse effect of Bt. cotton has so far been reported on parasites, predators and other non-target beneficial insects such as honey bees, silk worm and Lac worm. However some adverse effects of Bt. gene have been reported on Monarch butterfly, Lice wings (an insect predator) and tobacco horn worm.

(v) **Development of Resistance:** There is a fear that over a period of time the target insect may develop resistance to Bt. insecticidal gene. However, in cotton there is no report on development of resistance to target insect so far. The first report of development of resistance to Bt. toxin is that of diamond back moth, an important pest of *Brassica* crops the world over.

(vi) **Effect on Soil Ecology**: It is considered that Bt. crops will liberate some additional chemicals from their roots and have adverse effects on soil ecology. Bt. [cry1 Ac] gene has been used in many crops, but no adverse effects of this gene on soil ecology has been reported so far.

(vii) Evolution of New Viruses: There is fear that development of virus resistant cultivars may sometimes lead to development of new stronger variants of virus that can affect transgenic plants.

Summary

Bio-safety refers to the safe application of modern biotechnology with regard to human health, animal health and environment. Brief history and need for Cartagena bio-safety protocol are given. By June 2001, the Protocol had received 103 signatures. The membership is open to all countries of the World. The Protocol entered into force on 11 September 2003. The membership has been signed by 162 countries till November 2011. India signed the Cartagena Protocol in January 2001. However, countries like US, Argentina and Canada that produce about 90 percent of genetically engineered [GE] crops in the world have not ratified the Protocol, and are actively working to undermine it.

The governing body of the Protocol is the Conference of the Parties to the Convention serving as the meeting of the Parties to the Protocol (COP-MOP). The main function of this body is to review the implementation of the Protocol and make decisions necessary to promote its effective operation. The secretariat of the protocol is the same as for convention on biological diversity [CBD]. It is located at Montreal in Canada. Funds for the protocol are voluntarily contributed by contracting parties, developed country parties and other countries and sources. The objective and main elements of Cartagena bio-safety protocol are briefly presented.

There are two broad categories of risk from genetically modified organisms and their products, *viz*. (i) risk to human and animal health, and (ii) risk to the environment. Risk to human health includes toxicity, allergenicity, antibiotic resistance and nutritional uptake. The risk to environment include development of super weeds, loss of biodiversity, contamination of non-transgenic cultivars, effect on non-target organisms, development of resistance, effect on soil ecology and evolution of new viruses.

Questions

1. **What is Cartagena protocol? Explain briefly its history and objectives.**

2. **Describe in brief functions of Cartagena protocol.**

3. **Write short notes on the following:**
   (i) Need for Cartagena protocol
   (ii) Member of the governing body of Cartagena protocol.

4. **What are the risk from GMOs to the human health and animal health?**

5. **Describe risk from GMOs to the environment.**

6. **Define the following terms:**

   (i) Bio-safety (ii) Allergenicity

   (iii) Pathogenecity (iv) Super weeds

Chapter 16

Bio-safety Regulations for Transgenic Plants

Introduction

India has a well defined regulatory mechanism for development and evaluation of genetically modified organisms [GMOs] and their products. The department of biotechnology [DBT] and the Ministry of Environment and Forests [MoEF] are the two apex regulatory bodies that look into bio-safety aspects of GMOs. Guidelines for safety have been issued by the Department of Biotechnology [DBT] in 1990 covering research in biotechnology, field trials and commercial applications. Presently, there are following six competent authorities for implementation of regulations and guidelines for development and evaluation of genetically modified organisms in the country.

1. Recombinant DNA Advisory Committee [RDAC],
2. Review Committee on Genetic manipulation [RCGM],
3. Genetic Engineering Approval Committee [GEAC]- an apex body,
4. Institutional Bio-safety Committee [IBSC] attached to every organization engaged in recombinant DNA research,
5. State Bio-safety Coordination Committees[SBCC}, and
6. District Level Committees [DLC].

A brief description of the activities of the above committees is presented below.

1. Recombinant DNA Advisory Committee [RDAC]

This committee is constituted by the Department of Biotechnology. It takes note of development in biotechnology at national and International levels. RDAC prepares recommendations from time to time that are suitable for implementation for upholding the safety regulations in research and applications of GMOs and their products.

This committee prepared the Recombinant DNA Bio-safety Guidelines in 1990, which was adopted by the Government of India for conducting research and handling of GMOs in India.

2. Review Committee On Genetic Manipulation [RCGM]

The RCGM is constituted by DBT for a period of three years. It consists of chairman, co-chairman, member secretary, co-member secretary and 18-20 members from different research organizations. The RCGM works under the Department of Biotechnology and has the following functions.

- **(i)** To bring out manuals of guidelines specifying producers for regulatory process on GMOs in research, use and applications including industry with a view to ensure environmental safety.
- **(ii)** To review all ongoing r-DNA projects involving high risk, category and controlled field experiments.
- **(iii)** To lay down procedure for restriction or prohibition, production, sale, import and use of GMOs both for research and applications.
- **(iv)** To permit experiments with category II risks and above with appropriate containment.
- **(v)** To authorize imports of GMOs/transgenes for research purposes.
- **(vi)** To authorize field experiments in 20 acres in multi-location in one crop season with up to one acre at one site.
- **(vii)** To generate relevant data on transgenic material in appropriate systems.
- **(viii)** To undertake visits of sites of experimental facilities periodically, where projects with biohazard potentials are being pursued and also at a time prior to the commencement of the activity to ensure that adequate safety measures are taken as per the guidelines.

3. Genetic Engineering Approval Committee [GEAC]

It functions as a body under Ministry of Environment and Forests and is responsible for approval of activities involving large scale use of hazardous micro-organisms and recombinant products in research and industrial production from the environment angle. Its main functions are as follows:

- **(i)** To permit the use of GMOs and their products for commercial application.
- **(ii)** To adopt producers for production, sale, import and use of GMOs both for research and applications under Environment Protection Authority [EPA].
- **(iii)** To authorize large scale production and release of GMOs and their products into the environment.
- **(iv)** To authorize agencies or persons to have powers to take punitive action under the EPA.

4. Institutional Bio-Safety Committee [IBSC]

The IBSC is the nodal point for interaction with institution for implementation of the guidelines. It is constituted before undertaking any project involving manipulation of micro-organisms, plants or animals. It consists of Head of institution, 3 or more scientists working in molecular biology, one outside expert in the discipline and one member nominated by DBT. The main activities of IBSC are as follows:

(i) To note and approve r-DNA work.

(ii) To ensure adherence of r-DNA safety guidelines of government.

(iii) To prepare emergency plan according to guidelines.

(iv) To recommend to RCGM about category II risk or above experiments and to seek approval of RCGM.

(v) To act as a nodal point for interaction with statutory bodies.

(vi) To ensure experimentation at designated locations, taking into account approved protocol.

5. State Bio-Safety Coordination Committees [SBCC]

This committee is constituted in each state where research and application of GMO is carried out. It is headed by Chief Secretary of the State. He can nominate State Government representative in the activities related to field inspection of GMOs. Main functions of this committee are as follows:

(i) To ensure adherence to r-DNA safety guidelines issued by the government.

(ii) To review periodically the safety and control measures in various institutions handling GMOs.

(iii) To act as nodal agency at State level to assess the damage, if any, due to release of GMOs and to take on site control measure.

(iv) To coordinate activities related to GMOs in the state with central Ministries.

6. District Level Committees [DLC]

This is constituted at the district level and is the smallest committee. It is headed by the District Collector. Its functions are as follows:

(i) To monitor the safety regulations in the institutions.

(ii) It has powers to inspect, investigate and report to the SBCC or GEAC about compliance or non compliance of r-DNA guidelines or violations under EPA.

(iii) To act as nodal agency at District level to assess the damage, if any, due to release of GMOs and to take on site control measures.

Summary

In India, there are six competent authorities for implementation of bio-safety egulations and guidelines. These are: Recombinant DNA Advisory Committee[RDAC], Review Committee on Genetic manipulation [RCGM], Genetic Engineering Approval

Committee [GEAC]- an apex body, Institutional Bio-safety Committee [IBSC] attached to every institution engaged in recombinant DNA research, State Bio-safety Coordination Committees[SBCC}, and District Level Committees [DLC]. Roles of all these committees have been discussed.

Questions

1. **Discuss briefly bio-safety regulations related to GMOs in India. Describe any one of them in detail.**

2. **Explain briefly the role of Recombinant DNA Advisory Committee [RDAC].**

3. **Discuss in brief the constitution and functions of Review Committee on Genetic manipulation [RCGM].**

4. **Describe briefly the constitution and functions of Genetic Engineering Approval Committee [GEAC].**

5. **Describe briefly the functions of the following committees"**

   (i) Institutional Bio-safety Committee [IBSC]

   (ii) State Bio-safety Coordination Committees [SBCC]

   (iii) District Level Committees [DLC]

Section V

Miscellaneous Topics

Chapter 17

Plant Quarantine Order, 2003 (Regulation of Import into India)

Introduction

The Plant Quarantine order, 2003 was approved by the government of India, Ministry of Agriculture (Department of Agriculture and Cooperation) on 18th[h] November, 2003. This order may be called the Plant Quarantine (Regulation of Import into India) Order, 2003. This order has replaced the Plants, Fruits and Seeds order, 1989. This Order has been enacted for the purpose of prohibiting and regulating the import into India of agricultural articles. This order consists of seven chapters. Chapter 1 deals with title, scope and definitions of various terms. Chapter 2 deals with general conditions of import of seed, fruits and plants. Chapter 3 explains special conditions of import. Chapter 4 deals with post entry quarantine, Chapter 5 deals with appeals and revision. Chapter 6 explains powers of relaxation and Chapter 7 deals with repeal and savings. These are briefly presented as follows:

1. Scope and Coverage

This order may be called the Plant Quarantine (Regulation of Import into India) Order, 2003. This Order comes into force with effect from 1st January, 2004. It extends to whole of India

2. General Conditions for Import

(1) **Valid Permit**: As per this order, no consignment of plants and plant products and other regulated articles shall be imported into India without a valid permit issued by competent authority.

(2) **Materials for Import:** Plant materials of specified categories and species can be imported from specified countries only. Commercial import of consignments of seeds of coarse cereals, pulses, oil seeds and fodder seeds and seeds/stock material of fruit plant species for propagation shall only be permitted based on the recommendations of EXIM Committee of Department of Agriculture and Cooperation, except the specified trial material.

(3) **Application:** Every application for a permit shall be made at least one month in advance to the Issuing Authority, in Form PQ 01 for the import of plants and plant products for consumption and processing and in form PQ 02 for import of seeds and plants for propagation.

(4) **Fees:** A fee of Rs.150/- shall be payable along with the application for the import of seeds, fruits and plants for consumption and Rs.300/- for application for the import of seeds and plants for sowing or planting and the fee shall be payable in the form of Demand Draft payable to the Competent Authority having jurisdiction.

(5) **Issue of Permit:** The Issuing Authority shall issue permit in quadruplicate in form PQ 03 for import of plants and plant products for consumption and in form PQ 04 for import of seeds and plants for sowing or planting, if he is satisfied that the applicant meets all the necessary conditions. One copy of import permit shall be forwarded to the exporter in advance to facilitate incorporation of import permit number in the phytosanitary certificate issued by the exporting country. The import permit shall be issued for import of specified material from specified countries only.

(6) **Period of Validity:** The import permit issued shall be valid for a period of six months from the data of issue and valid for successive shipment provided the exporter and importer, bill of entry, country of origin and phytosanitary certificate are the same for the entire consignment. The issuing authority may, on request, extend the period of validity for a further period of six months after charging Rs. 200/- and Rs. 100/- as revalidation fee for propagation and consumption plant material respectively provided such request for extension of validity is made to the issuing authority before the expiry of the permit with adequate reasons to be recorded in writing.

(7) **Quantity for Import:** The quantity mentioned in the import permit if exceeds by up to 10 per cent may be allowed by charging additional inspection fee and import permit fee provided the excess quantity reflected in the phytosanitary certificate from the country of exporting. The import permit will become invalid if quantity exceeds more than 10 per cent of the quantity of import permit. Suppression of the facts or any material information while issue of import permit is liable to be cancelled or withdrawn.

(8) **Transfer:** The import permit issued shall not be transferable and no amendments to the permit shall be issued except for change of point of entry subject to reasons to be recorded in writing.

(9) Tag Colour: An orange and green color tag shall be issued in form PQ 05 in the case of permits issued for import of seeds and plants for sowing or planting so as to facilitate the identification of consignments at the time of their arrival at the point of entry.

(10) Contamination: No consignment of seed or grain shall be permitted to be imported with contamination of quarantine weeds. Every application for quarantine inspection and clearance shall be made in Form PQ 15.

(11) Places of Import: All the consignments of plants and plant products and other regulated articles shall be imported into India only through ports of entry as specified in Schedule-I and Inland Container, Depots/Container Freight Stations and foreign post offices falling within the jurisdiction of concerned plant quarantine station operating here under or those notified by the Government from time to time in this behalf.

All consignments of seeds and plants for propagation and regulated articles such as live insects, microbial cultures, bio-control agents and soil shall only be imported into India through regional plant quarantine stations of Amritsar, Chennai, Kolkata, Mumbai or New Delhi or through any other points of entry as may be notified from time to time for this purpose. However, no import of germplasm/transgenic plant material and genetically modified organisms shall be permitted through New Delhi Airport.

(12) Inspection of material: On arrival, at the first point of entry the consignment shall be inspected by the Plant Protection Adviser or any other officer duly authorized by him in this behalf and appropriate samples shall be drawn for laboratory testing, in accordance with the guidelines issued by Plant Protection Adviser from time to time.

(13) Quarantine Clearance: The Plant Protection Adviser or the officer authorized by him may, after inspection and laboratory testing, fumigation, disinfection or disinfestation, as may be considered necessary by him, accord quarantine clearance for the entry of a consignment or grant provisional clearance for growing under post-entry quarantine, as the case may be in form PQ 16 and or order deportation or destruction of the consignment in form PQ 17 in the event of non-compliance with the restrictions and conditions specified in this Order.

(14) Cost of Fumigation, etc: Where fumigation or disinfestation or disinfection is considered necessary in respect of a consignment of plants, seeds and fruits the importer shall on his own and at his cost arrange for the fumigation, disinfection or disinfestation of the consignment, through an agency approved by the Plant Protection Adviser under the supervision of an officer duly authorized by the Plant Protection Adviser in that behalf.

(15) Responsibilities of Importer: The importer of seed material has various responsibilities. Importer himself or his authorized agent has following responsibilities:

(i) To file an application for the quarantine inspection of imported seeds, plants and plant products or other regulated articles in the form PQ 15 along with copies of relevant documents and fees as prescribed under Schedule-IX payable by a demand draft to the competent authority.

(ii) To provide information on any plant and plant product and other articles covered under this Order and which are imported by him/her or are in his/her possession, to Plant Protection Adviser or any officer duly authorized by him.

(iii) To bring the consignments to the concerned plant quarantine station or to place of inspection, fumigation or treatment as directed by Plant Protection Adviser or any officer duly authorized by him.

(iv) To permit drawing of appropriate samples for inspection and laboratory investigation and extend necessary facilities towards the same.

(v) To open, repack and load into or unload from the fumigation chamber and seal the consignment.

(vi) To remove them after inspection and treatment according to the directions issued by the Plant Protection Adviser or any officer authorized by him.

(vii) To arrange deportation or destruction of the consignment at the cost of importer as may be deemed necessary by Plant Protection Adviser or an officer authorized by him.

(16) Container: The container carrying plants and plant products for import from other countries should be packed properly. It should not permit spillage of material or contamination with soil or escape of any pest. Moreover, the package or container shall not be opened or seals are broken anywhere in India.

(17) Phytosanitary Certificate: No consignment shall be permitted import unless accompanied by a Phytosanitary Certificate issued by an authorized officer at the country of origin in the form PQ 21 or at the country of re-export in form PQ 22; However, cut flowers, garlands, bouquets, dry fruits/nuts etc., weighing not more than two kilograms imported for personal consumption may be allowed to be imported without a Phytosanitary Certificate or an import permit.

(18) Packaging: The packed material that is appropriately treated permitted for import. The treatments shall include heat-kiln treatment at 560 C for a minimum of 30 hrs or Methyl Bromide fumigation at 48 g/cum for 32 hours or chemical impregnation of wood with wood preservatives such as copper chrome arsenic or any other approved treatment as per international standards and the treatment shall be endorsed in phytosanitary certificate. No article packed with packaging materials shall be released by the proper officers of customs unless the consignment is accompanied by a phytosanitary certificate in respect of said packing material; If no phytosanitary certificate is furnished in respect of said packaging material, the proper officer of customs shall grant out of charge only after clearance is obtained from local plant quarantine authorities, who shall grant clearance from the quarantine angle and may, if deemed fit, subject the said packaging material to treatment at the expense of importer.

(19) Import of soil, etc.: No import of soil, earth, compost, sand, plant debris along with plants, fruits and seeds shall be permitted except under the following conditions:

(i) The consignments of soil, earth, clay and similar material for any microbiological, soil mechanics, or mineralogical investigations and peat for horticultural purposes may be permitted through specified air or sea ports or land custom station, on applications made for that purpose.

(ii) The application for the purpose referred to in (i) above shall be made to the Plant Protection Adviser, at least one month in advance, in form PQ 06 along with a registration fee of Rs. 200/- by a bank draft drawn in favor of Accounts Officer, Directorate of Plant Protection, Quarantine and Storage, Faridabad.

(iii) The Plant Protection Adviser may, after scrutiny of the application, and if satisfied of the purpose, for which such consignment is being imported, issue special permit in Form PQ.

(iv) The consignments soil, peat or sphagnum moss etc., shall be inspected, fumigated, disinfected or disinfested by the importer from an agency approved by the Plant Protection Adviser under the supervision of an officer duly authorized by Plant Protection Adviser.

(20) Fees for inspection, fumigation, etc.: The importer of the consignment or his agent shall pay to the Plant Protection Adviser or any other officer duly authorized by him in this behalf, the fees prescribed for inspection, fumigation, disinfestation, disinfection of consignment.

(21) Permits required for import of Germplasm, Transgenic or Genetically Modified Organisms:

(i) No consignment of germplasm/transgenics/Genetically Modified Organisms (GMOs) shall be imported into India for research/ experimental purpose without valid permit issued by the Director, National Bureau of Plant Genetic Resources, New Delhi.

(ii) Every application for import of plant germplasm/transgenics/ genetically modified organisms for research/experimental purpose by the public/private organizations will be made to the Director, National Bureau of Plant Genetic Resources, New Delhi in form PQ 08 and the permit shall be issued in form PQ 09 in triplicate and a red/green tag in PQ 10 for germplasm and a Red/White tag in PQ 11 for transgenic/ Genetically Modified Organisms. Such permits for import of transgenic/ Genetically Modified Organisms shall be issued subject to the approval of Review Committee on Genetic Manipulation (RCGM) set- up by Department of Biotechnology.

(iii) No imported consignments of plant germplasm/transgenics/genetically modified pests shall be opened at the point of entry and it shall be forwarded to the Director, National Bureau of Plant Genetic Resources, New Delhi.

(22) Permit required for import of live insects and microbial cultures:

(i) No Consignment of live insects, microbial cultures or bio-control agents shall be permitted into India without valid import permit issued by the Plant Protection Adviser.

(ii) Every application for permit to import insects or microbial cultures including algae or bio-control agents, shall be made in the form PQ12 at least two months in advance to Plant Protection Adviser along with a fee of Rs. 200/- towards registration in the form of bank draft issued in favor of the Accounts Officer, Directorate of Plant Protection Quarantine and Storage, Faridabad.

(iii) The Plant Protection Adviser shall issue the permit in Form PQ13 in triplicate, if satisfied of the purpose for which import is made and subject to such conditions imposed thereon. A yellow- green color tag or label in the form PQ14 shall be issued which shall be affixed on the parcel at the time of export.

(iv) All the consignments of insects, microbial cultures and bio-control agents shall be permitted only through specified points of entry. The consignment of beneficial insects shall be accompanied by a certificate issued by National Plant Protection Organization at the country of origin with additional declarations for freedom from specified parasites and parasitoids and the bio-control agents free from hyper-parasites. The consignment of beneficial insects/bio-control agents shall be subjected to post-entry quarantine as may be prescribed by the Plant Protection Adviser.

(23) Permit required for import of plants and plant products

(i) No consignment of plants and plant products, if found infested or infected with a quarantine pest or contaminated with noxious weed species shall be permitted to be imported.

(ii) Every vessel carrying out bulk shipment of grains shall be inspected on board by an officer duly authorized by Plant Protection Adviser before the same accorded permission to offload the grain at the notified port of entry. On inspection, if found free from quarantine pests and noxious weed species, permission shall be accorded to off- load the grain at the port or order fumigation/treatment of grain on board or immediately upon unloading at the port, as the case may be, before such permission is granted for movement outside the port and subject to such conditions as imposed thereon.

(iii) The bulk shipment (s) of transgenic plants or plant products or genetically modified organisms shall be dealt as per the provisions of the Rules for manufacture, use, import, export and storage of hazardous micro-organisms,

(24) Permit for import of timber:

(i) No consignment of timber shall be permitted import unless the following conditions and requirements are fulfilled.

(a) The timber shall be stripped off its bark, either be squared or rounded and accompanied by an official statement that the wood has been appropriately fumigated/treated and such treatment shall be endorsed in the phytosanitary certificate issued at the country of origin or re-export, as the case may be.

(b) The timber shall be marked 'kiln-dried' or with any internationally recognized mark.

(ii) All the consignments of timber shall be inspected on board prior to unloading at the port of arrival by an officer duly authorized by PPA and, if necessary, fumigated/treated on board before unloading.

However, the above conditions shall not apply for containerized cargo, which shall be inspected by an authorized PQ officer after unloading of the containers from the ship at the port/container freight station or Inland Container Depots under the jurisdiction of concerned Plant Quarantine Station.

3. Special Conditions of Import

General conditions shall apply to import of all plant materials. In addition to the general conditions listed above, there are some specific conditions for import permit issued by an appropriate issuing authority.

Every consignment of plant species shall be accompanied by a Phytosanitary Certificate. The certificate is issued by the authorized officer at country of origin or by the country of re-export along with attested copy of original phytosanitary certificate. It has to declare that material is free from specified pests or that the specified pests do not occur in the country or state of origin. It should be supported by documentary evidence.

4. Post-Entry Quarantine

Plants and seeds, which require post-entry quarantine shall be grown in post-entry quarantine facilities duly established by importer at his cost, approved and certified by the Inspection Authority as per the guidelines prescribed by the Plant Protection Adviser. The period for which, and the conditions under which, the plants and seeds shall be grown in such facilities shall be specified in the permit.

(i) PET shall not apply to the import of tissue-cultured plants that are certified virus-free, but such plants, shall be subjected to inspection at the point of entry to ensure that the phytosanitary requirements are met with.

(ii) Every application for certification of post-entry quarantine facilities shall be submitted to the inspection authority in Form PQ 18. The inspection authority if satisfied after necessary inspection and verification of facilities shall issue a certificate in Form PQ 19.

(iii) At the time of arrival of the consignment, the importer shall produce this certificate before the Officer-in-Charge of the Quarantine Station at the entry point along with an undertaking in form PQ 20.

(iv) If the Officer- in-Charge of the Quarantine Station, after inspection of the consignment is satisfied, he shall accord quarantine clearance with post-entry quarantine condition on the production, by an importer, of a certificate from the inspection authority with the stipulation that the plants shall be grown in such post-entry quarantine facility for the period specified in the import permit.

(v) After according quarantine clearance with post-entry quarantine conditions to the consignments of plants and seeds requiring post-entry quarantine, the Officer-in-Charge of the Quarantine Station at the entry point shall inform the inspection authority, having jurisdiction over the post-entry quarantine facility, of their arrival at the location where such plants would be grown by the importer.

(vi) **Responsibilities of Importer:** The importer or his authorized agent shall be responsible for the following.

(a) To intimate the inspection authority in advance about the date of planting of the imported plant or seed.

(b) Not to transfer or part with or dispose the consignment during the pendency of post entry quarantine except in accordance with a written approval of inspection authority.

(c) To permit the inspection authority complete access to the post-entry quarantine facility at all times and abide by the instructions of such inspection authority.

(d) To maintain an inspection kit containing all requisite items to facilitate nursery inspection and ensure proper plant protection and upkeep of nursery records.

(e) To extend necessary facilities to the inspection authority during his visit to the nursery and arrange destruction of any part or whole of plant population when ordered by him in the event of infection or infestation by a quarantine pest, in a manner specified by him.

(vii) The inspection authority of concerned area of jurisdiction or any officer authorized by the Plant Protection Adviser in this behalf, in association with a team of experts shall inspect the plants grown in the approved post-entry quarantine facility at such intervals as may be considered necessary in accordance with the guidelines issued by the Plant Protection Adviser, with a view to detect any pests and advise necessary phytosanitary measures to contain the pests.

(viii) The inspection authority shall permit the release of plants from post-entry quarantine, if they are found to be free from pests and diseases for the period specified in the permit for importation.

(ix) Where the plants in the post-entry quarantine are found to be affected by pests and diseases during the specified period the inspection authority shall:

(a) Order the destruction of the affected consignment of whole or a part of the plant population in the post-entry quarantine if the pest or disease is exotic, or

(b) Advise the importer about the curative measures to be taken to the extent necessary, if the pest or disease is not exotic and permit the release of the affected population from the post-entry quarantine only after curative measures have been observed to be successful. Otherwise, the plants shall be ordered to be destroyed.

(x) Where destruction of any plant population is ordered by the inspection authority, the importer shall destroy the same in the manner as may be directed by the inspection authority and under his supervision

(xi) At the end of final inspection, the inspection authority shall forward a copy of the report of post-entry quarantine inspection duly signed by him to the Plant Protection Adviser under intimation to officer- in-charge of concerned plant quarantine station.

(xii) The importer shall be liable to pay the prescribed fee for inspection of plants in the Post-entry Quarantine facility.

5. Appeal and Revision

Appeal: If an importer is aggrieved by the decision of the inspection authority regarding the destruction of any plant population, he may appeal to the Plant Protection Adviser within 7 days from the date of communication of the decision giving the grounds of appeal.

(i) It shall be lawful for the Plant Protection Adviser to rely on the observations of the inspection authority and such other expert opinion, as he may deem necessary, for deciding the appeal.

(ii) The memorandum of appeal shall be accompanied by a bank draft in favor of the Plant Protection Adviser and payable at Faridabad, evidencing the payment of fee of Rs. 100/-

Revision: The Plant Protection Adviser may, at any time, call for the records relating to any case pending before the inspection authority for the purpose of satisfying itself as to the legality or propriety of any decision passed by that authority and may pass such order in relation thereto, as it thinks fit: But no such order shall be passed after the expiry of three months from the date of the decision; Moreover, the Plant Protection Adviser shall not pass any order prejudicial to any person, without giving him a reasonable opportunity of being heard.

6. Power of Relaxation

There are certain conditions under which Relaxation of Import Permit and Phytosanitary Certificate is allowed as given below:

(i) The Central Government may, in public interest, relax any of the conditions of this Order relating to the import permit and the phytosanitary certificate in relation to the import of any consignment. The Joint Secretary in-charge of Plant Protection in the Department of Agriculture and Cooperation shall be the competent authority for according the relaxation.

(ii) In the event of grant of relaxation by competent authority, the consignment shall be released after charging the fee for import permit and fee for plant quarantine inspection at five times of normal rates.

(iii) The provisions of this Order shall apply without prejudice to the Customs Act, 1962 or any other Acts or Order related to imports.

Comparison of Seed Order 1989 and Plant Quarantine Order, 2003

There are some similarities and some dissimilarities between seed order, 1989 and plan quarantine order, 2003. A brief comparison of these two orders is presented in Table 17.1.

Table 17.1: Comparison of Seed Order, 1989 and Plant Quarantine Order, 2003

Sl.No.	*Particulars*	*Seed Order, 1989*	*Plant Quarantine Order, 2003*
1.	Permit	Compulsory	Compulsory
2.	Permit Fee for plants, fruits and seed used for consumption	Rs.50/	Rs. 150/
3.	Permit Fee for plants, fruits and seed used for planting	Rs. 100/	Rs. 300/
4.	Validity of permit	Six months	Six months
5.	Extension of permit	Up to six months	Six months
6.	Fees for extension of permit	Rs. 50/and Rs. 100/	Rs. 100/and Rs. 200/
7.	Form used for Issue of permit for consumption	Form C	PQ 03
8.	Form used for Issue of permit for planting	Form D	PQ 04
9.	Import of material through	Specified sea and airports	Specified sea and airports
10.	Post entry quarantine	Applicable	Applicable
11.	Fee for post entry quarantine inspection to be paid by	Importer	Importer
12.	Appeal and revision	Applicable	Applicable
13.	Power of relaxation	Vested with government	Vested with government

7. Repeal and Savings

Repeals and Savings: The following orders and notifications are hereby repealed,

(i) Rules for regulating the import of insects into India.

(ii) Rules for regulating the import of fungi into India.

(iii) Import of cotton into India.

(iv) Plants, Fruits and Seeds Order, 1989.

An import permit issued by any competent authority, which is in force immediately before the commencement of this Order and which is consistent with this Order, shall continue in force and all appointments made and fees levied under the repealed Rules, Regulations and Orders, and in force immediately before such commencement shall likewise continue in force and be deemed to be made or levied in pursuance of this Order until revoked.

Summary

The Plant Quarantine order, 2003 was approved by the government of India, Ministry of Agriculture (Department of Agriculture and Cooperation) on 18th^{h} November, 2003. This order may be called the Plant Quarantine (Regulation of Import into India) Order, 2003. This order has replaced the Plants, Fruits and Seeds order, 1989. This Order has been enacted for the purpose of prohibiting and regulating the import into India of agricultural articles. This order consists of seven chapters. Chapter 1 deals with title, scope and definitions of various terms. Chapter 2 deals with general conditions of import of seed, fruits and plants. Chapter 3 explains special conditions of import. Chapter 4 deals with post entry quarantine, Chapter 5 deals with appeals and revision. Chapter 6 explains powers of relaxation and Chapter 7 deals with repeal and savings.

There are some similarities and some dissimilarities between seed order, 1989 and plan quarantine order, 2003. A brief comparison of these two orders is presented.

Questions

1. **Explain title and scope of Plant Quarantine Order, 2003.**

2. **Describe briefly general conditions of import of seed, fruits and plants.**

3. **What are special conditions for import of seed, fruits and plants as per Plant Quarantine Order, 2003 ?**

4. **Discuss briefly procedure of post entry quarantine of seeds, fruits and plants.**

5. **Give a brief account of procedure of appeal related to import of seed, fruit and plants.**

6. **Give a brief comparison of seed order, 1989 and plant quarantine order, 2003.**

7. **Write short notes on the following:**

   (i) Inspection authority
   (ii) Plant protection advisor
   (iii) Power of relaxation
   (iv) Seed import

Chapter 18

Intellectual Property Rights

Introduction

The property refers to wealth or valuable things earned by a person. It is estimated in terms of land, house, garden, industries, animals, gold, silver, diamond, money etc. accumulated by a person. Each person wants to earn property for four main purposes as given below:

- **(i)** To get a better or higher status in the society, because of a person is measured in terms of property accumulated by him or her. On the basis of property, the person is called as rich, middle class or poor.
- **(ii)** To lead a better quality of life in the society.
- **(iii)** To get a name and fame in the society. In the society, rich people get more respect than others.
- **(iv)** For better financial security of life for himself and for his children.

Types of Properties

The various valuable belongings of a person are broadly classified into two groups, *viz.*, 1. immovable property, and 2. movable property. These are defined below:

Immovable Property

It refers to fixed types of properties which can not move from one place to other. Such properties include land, buildings [house, flat, bungalow, farm house] and gardens. Such properties are protected by laws of land in each country.

Movable Property

It refers to the property which can be easily shifted from one place to other. Such property includes animals, farm machines, furniture, fixtures, gold, silver, diamond, and money. Such properties are also protected by public laws.

Intellectual Property

There is third type of property what we call intellectual property. The product/ process/idea which is outcome of the brain of a person and can be used on commercial scale for benefit of human kind is called intellectual property. In other words, Intellectual property refers to creations of the mind: inventions, literary and artistic works, and symbols, names, images, and designs used on commercial scale. The main features of intellectual property are given below:

- **(i)** It is measured in terms of new ideas, processes, products, inventions and innovations developed by a person.
- **(ii)** It requires lot of intellectual inputs in terms of thinking, planning and fine tuning of new ideas/products/processes etc.
- **(iii)** It requires considerable amount of funds and other resources to develop new products/processes.
- **(iv)** The main problem with intellectual property is that it can be copied, reproduced and used by others resulting in loss of inventor. Hence protection of intellectual property is essential so that the inventor can derive maximum benefits from his invention.

The rights to intangible property that is the product of the human intellect are referred to as intellectual property rights. Intellectual property may be protected by copyright, trademark or patent. The holder of intellectual property rights is usually the person or persons who developed the product or the organization that funded it.

Forms of Intellectual Property Rights

The Intellectual Property Rights are broadly divided into two groups, *viz.*, Primary Rights and Sui-generis Rights. These are briefly discussed below:

- **A. Primary Rights:** Primary rights include,Copyrights, Patents, Trademarks, Trade Secret, Trade names, Domain names, Industrial Designs, Geographical Indications etc.
- **B. Sui-generis Rights**: Sui-generis refers to things of their own kind or things with unique characteristics. Such rights include Database Rights, Mask work, Plant Breeders Rights, Traditional Knowledge, Moral Rights and Supplementary Protection Certificate.

Table 18.1: Forms of Intellectual Property Rights/Laws

Sl.No.	*Primary Rights*	*Sui Generis Rights*
1.	Copyrights	Database Rights
2.	Patents	Mask work
3.	Trademarks	Plant Breeders'Rights
4.	Industrial Design Rights	Farmers'Rights
5.	Utility Models	Moral Rights
6.	Geographical Indication	Supplementary Protection Certificate
7.	Trade Secret	Indigenous Intellectual Property
8.	Related Rights	
9.	Trade Names	
10.	Domain Names	

A. Primary Rights

1. Copyrights

Copyright refers to a document which grants exclusive right to the author/creator to publish and sell literary or musical or artistic work. In other words, the exclusive rights to reproduce, sell and distribute a work, prepare derivative works and display the work publicly is referred to as copyright. It is the right of an author, artist, publisher etc to retain ownership of works and to produce or contract others to produce copies. It is also defined as the legal right granted to an author, composer, playwright, publisher, or distributor to exclusive publication, production, sale, or distribution of a literary, musical, dramatic, or artistic work is called copyright. Main features of copyrights are briefly presented below:

(i) It provides protection for a specific duration.

(ii) It is applicable in all countries.

(iii) The copyright holder has the rights to authorize others to use the protected work.

(iv) It is applicable to books, movies, music, paintings, photographs and software.

(v) The information cannot be reproduced as such without written permission. However, the information or idea can be used by any one.

2. Patent

Patent has been defined in various ways by different authors.Patent refers to a document granting an inventor sole rights to an invention. It is an official document which grants sole rights to the inventor for manufacturing and marketing his product/process/invention to derive benefits.In other words, patent is a set of exclusive rights granted by a state to a patentee (the inventor or assignee) for a fixed period of time. Main points are as follows:

(i) It is a legal document issued by a federal government that grants exclusive rights for the production, sale and profit from the invention of a product or process for a specific period of time. Patents also grant the right to prevent others from copying the invention.

(ii) An official license is granted by the Patent Office to issue exclusive right to an individual or business for production or sale of a specific invention.

(iii) It is a grant made by a government to an inventor, assuring the sole right to make, use, and sell the invention for a certain period of time. Many audio and video technologies are covered by patents.

(iv) It is a document that allows the patent owner to prevent others from making, using or selling the invention protected by the patent.

3. Trademark

A trademark or trade mark is a distinctive sign of some kind which is used by an individual, business organization or other legal entity to uniquely identify the source of its products and/or services to consumers, and to distinguish its products or services from those of other entities. Main features of trademarks are briefly presented below:

(i) A trademark can be a word, name, symbol, device or mark which is used to identify and distinguish the goods or services of one company from goods or services of another.

(ii) Trademark is used to identify its product and to distinguish them from others. It is the name of a product made by a particular person or company.

(iii) The period of protection for a trademark varies, but can generally be renewed indefinitely.

4. Industrial Design

An industrial design - or simply a design - is the ornamental or aesthetic aspect of an article produced by industry or handicraft; registration and renewals provide protection for, in most cases, up to 15 years. Main points about industrial designs are given below.

(i) Industrial design is an applied art whereby the aesthetics and usability of products may be improved for marketability and production. Industrial Designers often utilize 3D Industrial Design is the field of developing physical solutions to meet a particular need. These physical solutions might include products.

(ii) Industrial design refers to any original shape, picture, or some combination applied to a useful article of manufacture. In other words, it refers tothe design of the mass-produced products of our everyday environment, from sinks and furniture to computers.

(iii) Industrial design refers to the professional service of creating and developing concepts and specifications that optimize the function, value, and appearance of products and systems for the mutual benefit of both user and manufacturer [Industrial Design Society of America].

5. Utility Models

A utility model is an exclusive right granted for an invention, which allows the right holder to prevent others from commercially using the protected invention, without his authorization, for a limited period of time. In its basic definition, which may vary from one country (where such protection is available) to another, a utility model is similar to a patent. In fact, utility models are also referred to as "petty patents" or "innovation patents, minor patent and small patent." The important points about utility model are listed below:

(i) Utility model is an intellectual property rights to protect inventions.

(ii) Utility models are also known as petty patents, petty inventions, petty innovations, utility innovations, minor patents and small patent.

(iii) The rights conferred by utility models are similar to those granted by patent laws but has a shorter term.

(iv) They are registered with National Patent Office.

(v) They are more suited to incremental inventions.

6. Geographical Indication

The World Intellectual Property Organization (WIPO) helps in protection of Geographical Indications [GI]. About 170 geographical indications have been registered by Lisbon Agreement members up to 1997. Now this number would much more. GI has been defined in various ways. Some important definitions and features of GI are given below:

(i) This is the name of a region, a specific place or, in exceptional cases a country, used to describe an agricultural, natural or manufactured goods product or a food stuff.

(ii) A geographical indication is a sign used on goods that have a specific geographical origin and often possess qualities or a reputation that are due to that place of origin. In other words, a geographical indication is a name or sign used on certain products or which corresponds to a specific geographical location or origin (*e.g.* a town, region, or country).

(iii) So far geographical indications have been registered in some advanced countries because they have developed the system of protecting GI.

(iv) In the past, wines,spirits, cheeses, tobacco, which account for 88 per cent, have been registered as GI. Wines and spirits account for almost 71 per cent of all registrations.

(v) In addition to wines and spirits, there is need for protection of agricultural products as GI. For this there is need to establish a multilateral system of identification and registration of GI.

(vi) Geographical Indication may be an agricultural, natural or manufactured goods or product.

7. Trade Secret

A "trade secret" is anything [a formula, process, method, mechanism, tool, pattern or device] which the disclosing party desires to keep secret. Trade secrets usually include such things as the manufacturing details for a product, variations or alternative uses. The main points about trade secret are briefly presented as follows:

- **(i)** There is no specific period for trade secret. It may continue life long or for generations together.
- **(ii)** There is no need of registration for trade secret.
- **(iii)** It does not provide opportunity to others forimprovement of innovation.
- **(iv)** It is not applicable to books, equipments, plant varieties, designs which are openly used.

8. Related Rights

Related rights is a term in copyright law, used in opposition to the term "authors' rights". The term neighbouring rights is exactly equivalent, and a more literal translation of the original French *droits voisins* Related rights in civil law are rights which are similar to authors' right but which are not connected with the actual author of the work. Both authors' rights and related rights are copyrights in the sense of English or U.S. law. Main points about related rights are given below:

- **(i)** Related rights are similar to copyrights.
- **(ii)** Related rights come under primary property rights.
- **(iii)** Related rights are also known as neighouring rights.
- **(iv)** Related rights are closer to copyrights but are not covered by the Berne Convention.
- **(v)** Related Rights are covered by Rome Convention, 1961.

9. Trade Names

A trade name, also known as a trading name or a business name, is the name which a business trades under for commercial purposes, although its registered, legal name, used for contracts and other formal situations, may be another. Pharmaceuticals also have trade names (*e.g.* "Aspirin"), often dissimilar to their chemical names ("acetylsalicylic acid"). Trading names are sometimes registered as trademarks or are regarded as brands. A trade name is the name under which a product is commercially known. The commercial name by which a chemical is known is called trade name. One chemical may have a variety of trade names depending on the manufacturers or distributors involved Main points related to trade name are given below:

- **(i)** It is also known as business name, trading name, assumed name, brand name and corporate name.
- **(ii)** It may or may not be registered
- **(iii)** It distinguishes a particular business from others.

(iv) It can be exclusive or non-exclusive.

(v) It is generally used on letter heads and bank accounts.

(vi) It is the name given to a particular substance by each company that manufactures it.

(vii) It is the name under which a company conducts its business.

(viii) It is s the business name of the person or organization making and/or selling a particular product. "General Motors, Inc" is the "trade name" of the company.

(ix) It is used by a company to describe and distinguish its brand of a generic product. Kleenex is a trade name for a brand of tissue; Xerox, a single brand of copier. A "trade name" is the business name of the person or organization making and/or selling a particular product. "General Motors, Inc" is the "trade name" of the company. Trade names are registered by the state in which the person or company is based.

10. Domain Names

Domain refers to the unique name that identifies an Internet site. Domain names have two or more parts separated by dots. For example www.ask-edi.com or www.bcbsks.com The term domain name has multiple related meanings:

(i) It identifies a web site on the internet (*e.g.*, company.com); also referred to as a URL.[Uniform Resource Locator]

(ii) It allows reference to Internet sites without knowing the true numerical address.

(iii) It refers to the official name of a computer connected to the Internet (*e.g.*, "www.microsoft.com").

(iv) It is the unique name of a computer on the Internet that distinguishes it from other systems around the world.

(v) They are sometimes colloquially (and incorrectly) referred to by marketers as "web addresses".

B. *Sui-Generis* Rights

Sui-generis is a Latin expression, literally meaning of its own kind/genus or unique in its characteristics. In law, it is term of art used to identify a legal classification that exists independently of other categorizations because of its uniqueness or due to the specific creation of an entitlement or obligation. In Intellectual Property there are rights which are known as being sui generic to owners of a small class of works. Such Intellectual Property Rights include (i) database Rights, (ii) mask work, (iii) plant breeders' rights, (iv) farmers' rights, (v) moral rights, (vi) supplementary protection certificate, and (vii) indigenous intellectual property, A brief account of all these is presented in this chapter.

1. Database Rights

Database rights are a form of exclusive right introduced by European Union Law in 1996 in those countries which follow EU Law. A 'database right' is an intellectual property right given to a computer database. The main points about database rights are as follows:

(i) In most countries database are covered by copyright law to some degree.

(ii) The database rights were introduced by European Union Law to provide greater protection to collection of information.

(iii) Database right is covered by *sui generic* rights.

(iv) **Duration**. Database rights last for 15 years, but can be extended if the database is updated.

(v) **Protection**. Database right prevents copying of substantial parts of database. The protection is over the information, not over the form of expression.

Table 18.2 List of Sui Generic Rights

Sl.No.	*Items of Sui Generic Right*	*Definition or Brief Description*
1.	Database rights	A 'database right' is an intellectual property right given to a computer database.
2.	Mask Work	A mask work is a two or three-dimensional layout of an integrated circuit (IC or "chip").
3.	Plant Breeders' Rights	Rights granted to the breeder of a new variety of plant.
4	Farmers' Rights	The legal rights provided to farmers to save, use, sow, replant, exchange, share or sell his farm produce.
5	Moral Rights	Moral rights are a special extension to copyright that gives the originator of a copyrighted work rights over its use.
6.	Supplementary protection certificate	A supplementary protection certificate (SPC) is a sui generis, patent-like, intellectual property right that is available for medicinal and plant protection products.
7.	Indigenous Intellectual Property	The knowledge, innovations and practices of indigenous people and local communities.

2. Mask Work Rights

A mask work is a two or three-dimensional layout of an integrated circuit (IC or "chip"), *i.e.* the arrangement on a chip of semiconductor devices such as transistors and passive electronic components such as resistors and interconnections. Main points related to mask work are listed as under:

(i) Mask work cannot be protected under copyright law or patent law.

(ii) The mask work exclusive rights were first granted in USA by Semiconductor Chip Protection Act of 1984. Such legislations also exist in Canada and Australia.

(iii) Duration: Rights for semiconductor mask work lasts only for two years (if unregistered) or 10 years (if registered).

(iv) The exclusive rights in mask work are somewhat like those of copyright: the right to reproduce the mask work or distribute an IC made using mask work.

(v) Reproduction for reverse engineering of a mask work is permitted by law.

(vi) Mask work rights exist when they are created, even without registration.

(vii) Protection under mask work cannot be provided to a mask that is not original.

Database and Copyright

There are some similarity and some differences between database and copyrights which are presented in Table 18.3.

Table 18.3: Comparison of Database and Copyrights

Sl.No.	*Particulars*	*Database Rights*	*Copyrights*
1 .	Governed by	Sui generic rights	Primary intellectual rights
2.	Registration	Not compulsory	Not compulsory
3.	Period of protection	15 years	70-120 years
4.	Protection over	Information	Expression
5.	Validity	Country of registration	Country of registration
6.	Enforcement	After creation	After creation

There are some similarity and some differences between database and mask work which are presented in Table 18-4.

Tanle 18.4: Comparison of Database and Mask Work

Sl.No.	*Particulars*	*Database Rights*	*Mask Work*
1.	Governed by	Sui generic rights	Sui generic rights
2.	Registration	Not compulsory	Not compulsory
3.	Period of protection	15 years	Unregistered 2 Registered 10 years
4.	Protection over	Information	Integrated Circuits
5.	Validity	Country of registration	Country of registration
6.	Enforcement	After creation	After creation

3. Plant Breeders' Rights

Plant breeders' rights, also known as plant variety rights (PVR), are intellectual property rights granted to the breeder of a new variety of plant. Plant Breeders Rights are granted to novel plant varieties that are distinctive, uniform, and stable (eg, cultivars bred true-to-type for desired traits). The legal protection of a new plant variety is granted to the breeder or his successor. The effect of PBR is that prior authorization is required before the material can be used for commercial purposes.

Main Features

Main features of Plant Breeders' Rights in relation to registration, duration, validity, matters covered, requirements, transfer, control, enforcement, rights protected and infringement. are briefly presented below:

(i) **Registration:** Registration is essential to get legal rights. There is no protection without registration. A nominal processing Fee is charged for registration.

(ii) **Duration:** The period of protection varies with plant species. It is 20 years for field crops, 25 years for trees, vines, fruit trees, ornamental trees and shrubs [UPOV 1991].

(iii) **Validity:** The protection of the Plant Breeders' Rights. is valid only in the country where it has been registered. The protection in other countries can be obtained by filling separate application in each country.

(iv) **Matters Covered:** The protection right can be granted for varieties of all botanical genera and species. The variety should have a designation [name] as per the rule of International Code of Nomenclature.

(v) **Requirements:** There are four basic requirements for protection of a variety under PBR, *viz.* (i) novelty, (ii) distinctiveness, (iii) uniformity and (iv) stability.

(vi) **Transfer:** An authorized plant breeder has right to authorize other interested persons for commercial production and marketing of his variety on royalty basis.

(vii) **Control:** The authorized plant breeder has right to prevent other from commercial production and marketing of his variety without permission.

(viii) The Plant Breeders' Rights come into force immediately after registration of the variety.

(ix) **Rights Protected:** It provides exclusive rights to the owner for commercial production and marketing of his variety.

(x) **Infringement:** Unauthorized production and marketing of a registered variety by other person amounts to infringement. The owner has the right to take legal action against the infringer and claim damages.

Advantages of PBR

1. Breeders get benefit of their variety.
2. PBR help in faster development of seed industry.
3. PBR lead to improvement in quality because of competition.
4. PBR are useful in procurement of good material on payment basis.
5. PBR help in enrichment of genetic resources.

Disadvantages of PBR

1. It will promote monopoly.
2. It will encourage unhealthy practices.
3. It may lead to increase in prices.
4. There will be reduction in genetic variability.
5. There will be compulsion to purchase frsh seed every year.

4. Farmers' Rights

Farmers' rights refer to the rights arising from the past, present and future contributions of farmers in conserving, improving and making available plant or animal genetic resources, particularly those in the centres of origin/diversity. In other words, the legal rights provided to farmers to save, use, sow, replant, exchange, share or sell his farm produce including seed of a variety protected under Plant Variety Protection Act refer to Farmers Rights. The purpose of these rights is to "ensure full benefits to farmers and support the continuation of their contributions.

The FAO Conference held in Rome from 11-29 November, 1989 endorsed the concept of Farmers Rights with a view to:

(i) Ensuring global recognition of the need for conservation and the availability of sufficient funds for these purposes;

(ii) Assisting farmers and farming communities throughout the world, especially those in areas of original diversity of plant genetic resources, in the protection and conservation of their PGR and of the natural biosphere; and,

(iii) Allowing the full participation of farmers, their communities and countries in the benefits derived, at present and in the future, from the improved use of PGR.

5. Moral Rights

Moral rights are rights of creators of copyrighted works generally recognized in civil law jurisdictions and first recognized in France and Germany, before they were included in the Berne Convention for the Protection of Literary and Artistic Works in 1928. Moral rights are a special extension to copyright that gives the originator of a copyrighted work rights over its use. The originator can specify moral conditions on the use or exploitation of their creation, even when the rights are licensed or sold to others'. The main points related to moral rights are presented below:

(i) These are rights provided to authors and directors of copyrights to protect their personal interest in relation to their creations. These rights are provided beyond those strictly recognized in copyright law.

(ii) Moral Rights were granted to authors under the Berne Convention (1989) giving the author (i) right to attribution, and (ii)) right to integrity.

(iii) The creator of the work has the right of attribution and integrity, even after sale or transfer of the copyright. In other words, the creator or author shall have the right to claim authorship of the work even after sale or transfer of the copyright.

(iv) Transfer: The copyrights can be transferred or assigned to others, but moral Right cannot be transferred or assigned.

(v) The author has right to object to any distortion, mutilation or other modification which would be prejudicial to his honour or reputation.

6. Supplementary Protection Certificate

This term is used in European Countries. This is an extension of the term given to pharmaceutical or plant protection patent. A supplementary protection certificate (SPC) is a sui generis, patent-like, intellectual property right. This type of right is available for medicinal products, such as drugs, and plant protection products, such as insecticides, and herbicides. Main points related to supplementary protection certificate are given below:

(i) It is in use in European Union member countries.

(ii) It is a sui generis, patent like, intellectual property right.

(iii) The SPC comes into force only after the expiry of corresponding patent.

(iv) Duration: It has maximum life time of 5 years when granted after 10 years after filing date of a patent. The term would be less than 5 years if it is granted after 5 years from the date of filing the corresponding patent. However, the market exclusivity cannot exceed 15 years.

(v) The SPC provides long term authorization to a patent for marketing.

(vi) Conditions of Issue: It is issued against a patent generally immediately on expiry of the term of the patent. It cannot be issued before 5 years from filing date of corresponding patent.

(vii) The registration is required which is issued on national basis. A separately application is to be filed for registration in each country.

7. Indigenous Property Rights

The Indigenous Knowledge also known as Traditional Knowledge [TK] refers to the knowledge, innovations and practices of indigenous people and local communities. In other words, it is the: knowledge gained through tradition from ancestors about plants and animals, medicines, agricultural practices and other matters. Main points about traditional knowledge are listed below:

(i) Traditional knowledge can be protected by *sui generic* rights.

(ii) Traditional knowledge is also known as indigenous knowledge or local knowledge or traditional ecological knowledge.

(iii) It covers the matured long-standing traditions and practices of certain regional, indigenous, or local communities. In other words, it is gained from experience gained over the centuries and adapted to local culture and environment.

(iv) It includes knowledge about plants, animals, indigenous treatment of diseases through herbal medicines, cultural practices,etc.

(v) Traditional knowledge is based on teachings and experiences and is orally passed on from generation to generation to some persons only. Earlier there were no written document on traditional knowledge.

(vi) Some forms of traditional knowledge are expressed through stories, legends, folklore, rituals, songs and even laws.

(vii) There is difference between local knowledge and traditional knowledge. The former existed from decade to century, whereas the latter existed from millennia.

(viii) It is collectively owned by the people or natives of an ecosystem or region.

(ix) Traditional knowledge incorporates knowledge of ecosystem and code of ethics governing appropriate use of environment. This code includes rules and conventions promoting desirable ecosystem relations, human-animal interactions and even social relationships. A combination of traditional and non-traditional knowledge will form a rich and distinctive understanding of life and the world.

The ecological knowledge is composed of both traditional knowledge and experiential knowledge[gained through personal experience.

Traditional knowledge is very important almost in every sphere of life. Hence, its protection is very necessary otherwise this valuable knowledge will be gradually lost.

Advantages of IPR

Intellectual Property Rights refer to the legal rights provided to an inventor to derive economic benefits from his invention/innovation. The main advantages of IPR are as follows:

1. It promotes healthy competition for invention/innovation among the intellectuals.
2. It helps in improving the quality of the product.
3. It makes available new ideas/technologies to different countries.
4. It leads to faster development of industries/organizations engaged in research and development work.

Disadvantages of IPR

There are some disadvantages of Intellectual Property Rights which are briefly presented below.

1. The procedure of registration, particularly of patents, is very lengthy.
2. It involves lot of money transaction in registration, renewal and licensing.
3. It invites lots of court cases due to infringements.
4. It may lead to monopoly of right holders, etc.

Summary

The property refers to wealth or valuable things earned by a person. It is estimated in terms of land, house, garden, industries, animals, gold, silver, diamond, money etc. accumulated by a person. The various valuable belongings of a person are broadly classified into two groups, *viz.*, 1. immovable property, and 2. movable property.

Immovable Property refers to fixed types of properties which can not move from one place to other. Such properties include land, buildings [house, flat, bungalow, farm house] and gardens. Such properties are protected by laws of land in each country.

Movable Property refers to the property which can be easily shifted from one place to other. Such property includes animals, farm machines, furniture, fixtures, gold, silver, diamond, and money. Such properties are also protected by public laws.

There is third type of property what we call intellectual property. The product/process/idea which is outcome of the brain of a person and can be used on commercial scale for benefit of human kind is called intellectual property.

The Intellectual Property Rights are broadly divided into two groups, *viz.*, Primary Rights and Sui-generis Rights. Primary rights include,Copyrights, Patents, Trademarks, Industrial designs rights, utility models, geographical indication, trade secret, related rights, trade name and domain name.

A brief account of primary property rights has been presented. Sui generic rights have been discussed in the chapter 13. Advantages and disadvantages of IPR have been discussed.

Sui-generis is a Latin expression, literally meaning of its own kind/genus or unique in its characteristics. In law, it is term of art used to identify a legal classification that exists independently of other categorizations because of its uniqueness or due to the specific creation of an entitlement or obligation. In Intellectual Property there are rights which are known as being sui generic to owners of a small class of works. Such Intellectual Property Rights include (i) database Rights, (ii) mask work, (iii) moral rights, (iv) supplementary protection certificate (v), plant breeders rights and (vii) traditional knowledge, important points related to all above items have been discussed.

Questions

1. **Define property. Describe briefly different type of properties.**

2. **What do the following terms signify?**

   (i) Immovable property　　(ii) Movable property

   (iii) Intellectual property　　(iv) *Sui-generis* rights

3. **Define the following terms:**

   (i) Copyright　　(ii) Patent

   (iii) Trademarks　　(iv) Trade Secret

   (v) Industrial design　　(vi) Geographical Indication

4. **Differentiate between the following:**
   (i) Movable and immovable property
   (ii) Copyright and patent
   (iii) Trademark and trade Secret
   (iv) Trade Name and Domain Name

5. **What is Intellectual Property? List different forms of Intellectual Properties and describe any one in detail.**

6. **Define exclusive rights. Explain briefly various types of primary rights.**

7. **Describe briefly advantages and disadvantages of Intellectual Property Rights.**

8. **Describe briefly the merits and demerits of copyrights, patents, trademark and trade secret.**

9. **Write short notes on any two of the following:**
   (i) Utility Models
   (ii) Related Rights
   (iii) Trade Name
   (iv) Domain Name

10. **Give in brief comparison of copyrights and patents.**

11. **Describe briefly trademarks and trade secret. Give their merits and demerits.**

12. **Describe briefly utility models and related rights.**

13. **Describe briefly trade names and domain names**

14. **What is sui generic right ? List various items covered by such rights and describe any one of them in detail.**

15. **Define database and describe it significance.**

16. **Explain briefly disadvantages of sui generic rights.**

17. **Write short notes on the following:**
   (i) Database rights
   (ii) Moral Rights
   (iii) Mask Work
   (iv) Traditional knowledge

18. **Define mask work and discuss its main features.**

19. **What do you mean by plant breeders' rights? Give merits and demerits of PBR.**

20. **Define traditional knowledge. Give its advantages and disadvantages.**

21. Give a brief account of supplementary protection certificate.

22. What do the following terms signify?

(i) *Sui generic* rights

(ii) Plant breeders' rights

(iii) Farmers' rights

(iv) Biodiversity

23. Give a brief comparison of moral rights and traditional knowledge.

24. Describe briefly Farmers' Rights and their significance.

Chapter 19

World Trade Organization

Introduction

The World Trade Organization [WTO] was established in 1995 as the successor to the General Agreement on Tariffs and Trade (GATT), which was itself established after the Second World War. The overall system was developed through a series of trade negotiations, or rounds, held under GATT. The WTO can be defined as an international organization that regulates trade and tariffs between nations in order to ensure that trade flows smoothly, predictable and as freely as possible. The main points are listed below.

1. It was created by the Uruguay Round of GATT to enforce free trade of manufactured goods, services, and intellectual property.
2. It is an international organization that monitors and enforces rules governing global trade.
3. It is designed to supervise and liberalize international trade.
4. It provides the ground rules for international trade and commerce.
5. It was created through a multilateral agreement on trade and investment.
6. It has been formed by the agreement of member nations.

Main Features

The discussion of World Trade Organization on the basis member countries, organizational structure, headquarters, objectives, functions, principles, materials covered for trading, advantages, disadvantages and effects is briefly presented below:

1. Member Countries

The membership is open to all countries. Interested countries become member of

WTO. India has joined membership of WTO. Till July 27, 2007, 151 countries have become member of WTO [See Appendix 2].

2. Organizational Structure of WTO

The World Trade Organization came into force on January 1, 1995, fully replacing the previous GATT Secretariat as the organization responsible for administering the international trade regime. The basic structure of the WTO includes the following bodies:

(i) The Ministerial Conference

This is composed of international trade ministers from all member countries. This is the governing body of the WTO, responsible for setting the strategic direction of the organization and making all final decisions on agreements under its wings. The Ministerial Conference meets at least once every two years. Although voting can take place, decisions are generally taken by consensus all member countries.

(ii) The General Council

This is composed of senior representatives (usually ambassador level) of all members. It is responsible for overseeing the day-to-day business and management of the WTO, and is based at the WTO headquarters in Geneva. In practice, this is the key decision-making arm of the WTO for most issues. Several of the bodies described below report directly to the General Council.

(iii) The Trade Policy Review Body

This is also composed of all the WTO members, and oversees the Trade Policy Review Mechanism, a product of the Uruguay Round. It periodically reviews the trade policies and practices of all member states. These reviews are intended to provide a general indication of how states are implementing their obligations, and to contribute to improved adherence by the WTO parties to their obligations.

(iv) The Dispute Settlement Body

This is also composed of all the WTO members. It oversees the implementation and effectiveness of the dispute resolution process for all WTO agreements, and the implementation of the decisions on WTO disputes. Disputes are heard and ruled on by dispute resolution panels chosen individually for each case, and the permanent Appellate Body that was established in 1994. Dispute resolution is mandatory and binding on all members. A final decision of the Appellate Body can only be reversed by a full consensus of the Dispute Settlement Body.

(v) The Councils on Trade in Goods and Trade in Services

These councils operate under the mandate of the General Council and are composed of all members. They provide a mechanism to oversee the details of the general and specific agreements on trade in goods (such as those on textiles and agriculture) and trade in services. There is also a Council for the Agreement on Trade-Related Aspects of Intellectual Property Rights, dealing with just that agreement and subject area.

(vi) The Secretariat and Director General

The Director General of the WTO reside in Geneva, in the old home of GATT. The Secretariat now numbers just under 550 people, undertakes the administrative functions of running all aspects of the organization. The Secretariat has no legal decision-making powers but provides vital services, and often advice, to those who do. The Secretariat is headed by the Director General, who is elected by the members.

(vii) The Committees

There are several committees, but two committees, *viz.* (a) committee on trade and development [CTD], and (b) Committee on Trade and Environment [CTE] are very important. They have specific mandates to focus on these relationships, which are especially relevant to how the WTO deals with sustainable development issues. The Committee on Trade and Development was established in 1965. The forerunner to the Committee on Trade and Environment (the Group on Environmental Measures and International Trade) was established in 1971, but did not meet until 1992. Both Committees are now active at discussion grounds but do not actually negotiate trade rules.

3. Headquarters

The World Trade organization is based at Geneva in Switzerland In other words, Geneva is the headquarters of WTO. The WTO is headed by the Director General who lives at Geneva in Switzerland in the old home of GATT.

4. Objectives

The main objective of World Trade Organization is to ensure smooth, free, fair and predictable global trade flow.

5. Functions

There are four main functions of World Trade Organization as given below:

(i) Administering trade agreements [enrolment of members].

(ii) Acting as a forum for trade negotiation.

(iii) Settling trade disputes and

(iv) Reviewing national trade policies to improve transparency and to assess their impact.

6. Principles

The WTO agreements are based on certain principles. Some important principles of WTO are briefly presented below:

(i) WTO rules have been framed by consensus among all member countries. The modification in any rule is also done by consensus among all member countries.

(ii) WTO permits non discriminating trading system for all member countries.

(iii) The membership is open for all countries of the world. Any country can become member of WTO.

(iv) In WTO, all countries have equal rights irrespective of their status.

(v) WTO functions purely on democratic pattern. All decisions are taken by consensus among member countries.

(vi) Each country gets guarantee for its export in the market of other member countries.

(vii) WTO is the government organization. The private sector can participate only in specific activities such as seminar and symposiums.

(viii) WTO does not force any country to become its member. Interested countries join WTO.

(ix) WTO agreements keep into account various issues such as environmental safety, job opportunities and development. It does not ignore economic development.

(x) WTO improves standard by enhancing per capita income of those engaged with WTO activities.

Materials Covered for Trading

WTO permits a variety of items for trading across the nation. The list of such items is presented in Table 19.1.

Table 19.1: List of Items Permitted for Trading by WTO

Sl.No.	*Group of Items*	*Actual Items*
1.	Agriculture	Cereals, pulses, oilseeds, oils, sugar etc.
2.	Horticulture	Fruits, vegetables, cut flowers etc.
3.	Textiles	Fibre, yarn, fabrics and garments.
4.	Automobiles	Cars, buses, tractors, trucks, road making machines and their spare parts.
5.	Electronic Goods	Television, computers, and laboratory instruments and equipments.
6.	Electrical goods	Refrigerators, electric iron, heaters, air coolers, deep fridges etc.
7.	Consumer Goods	Soaps, cosmetics, creams, tooth pastes, tooth brush etc.
8.	Household Goods	Utensils, furniture, sofa, bed, dinning table etc.
9.	Literary Goods	Books, papers, pens etc.
10.	Services	Banks, Insurance firms, phone companies, transport companies, hotels etc.
11.	Defense Items	Guns, rifles, fighter aircrafts, missiles, ships, helicopters etc.
12.	Intellectual Property	Copyrights, trademarks, industrial designs, Integrated circuit layout designs, trade secret geographical indications, etc.

Advantages of WTO

There are several advantages of WTO which are listed as follows:

(i) Improvement in the quality of product or service due to competition among companies.

(ii) There will be reduction in the price again due to competition among companies.

(iii) WTO provide job opportunities by creating new jobs.

(iv) WTO provides a wide range of variety in various items for choice.

(v) WTO improves relationship among trading countries due to frequent interactions.

(vi) WTO leads to faster development and growth of member countries by floating their goods in the international market.

(vii) Since each item is marketed by several countries, there will be no monopoly of any country.

(viii) WTO operates on democratic pattern, each country has equal rights and all decisions are taken with consensus of member countries.

(ix) WTO provides atmosphere for free and fair deal.It curbs corruption.

(x) WTO leads to increase in the income by permitting trade across the nations.

Disadvantages of WTO

There are two main disadvantages of WTO. Firstly it leads to several trade disputes and secondly, it has adverse effects on the growth of small industries.

Summary

The World Trade Organization [WTO] was established in 1995 as the successor to the General Agreement on Tariffs and Trade (GATT), which was itself established after the Second World War. The overall system was developed through a series of trade negotiations, or rounds, held under GATT. The WTO can be defined as an international organization that regulates trade and tariffs between nations in order to ensure that trade flows smoothly, predictable and as freely as possible.

The discussion of World Trade Organization on the basis member countries, organizational structure, headquarters, objectives, functions, principles, materials covered for trading, advantages and disadvantages is briefly presented.

The WTO consists of the Ministerial Conference, the General Council, Trade Policy Review Body, Dispute Settlement Body, Councils on Trade in Goods and Trade in Services, Secretariat and Director General and various committees. The role of all these have been explained. Till July 2007, 151 countries have joined WTO. The main objective of World Trade Organization is to ensure smooth, free, fair and predictable global trade flow.

Questions

1. **Define WTO describe briefly its organizational structure.**

2. **Discuss briefly the materials covered for trading under WTO.**

3. **Describe briefly principle of WTO.**

4. Explain briefly Advantages of WTO.

5. Write short notes on the following

(i) Headquarters of WTO
(ii) Objectives of WTO
(iii) Functions of WTO
(iv) Demerits of WTO

6. Describe briefly WTO on the basis of following:

(i) Membership
(ii) Objectives
(iii) Dispute settlement
(iv) Materials covered

7. Describe briefly the requirements for registration of a variety under PPV and FR Act 2001.

8. Describe the role of following councils of WTO:

(i) General council
(ii) Goods Council
(iii) Service Council
(iv) Intellectual property council

9. Give full form of the following:

(i) WTO
(ii) GATT
(iii) TPRB
(iv) TPRM

10. Describe briefly the impact of WTO on the following:

(i) Quality of product
(ii) Price of the product
(iii) Choice of items
(iv) Job opportunities

Chapter 20

Globalization

Introduction

Globalization is the process of organizing the whole world into a single integrated marketing unit. It is also defined as the process of trans border free flow of products, services, people, culture, technology, and finance. It leads to the integration of economic, cultural, political, and social systems across national borders Globalization is also referred to as internationalization by some persons Both these terms are used as synonyms. However, some people use these terms separately. This chapter deals with various globalization such as phases of globalization, types [kinds] of globalization, global connectivity, measurement of globalization, factors affecting globalization, advantages of globalization and disadvantages of globalization.

1. Phases of Globalization

Globalization is not a new phenomenon. It started with human civilization. In the past 130 years, modern historians have identified three stages or phases of globalization, *viz.*, first phase, second phase and third phase. These are briefly discussed below:

(a) First Phase

This phase of globalization started from 1870 and ended in 1913 with outbreak of the First World war. The main features of first phase are given below.

- **(i)** There was marked mobility of capital.
- **(ii)** The labour mobility was high.
- **(iii)** The magnitude of free trade was limited.
- **(iv)** The global institutions were non existent.
- **(v)** The National Institutions were heterogeneous and were not well organized.

The First World War had adverse effects on the process of globalization *i.e.* flow of products, services, labour, and technology across the countries. It started resuming since 1930 again and got setback from 1940 to 1944 due to second World War.

(b) Second Phase

The second phase of globalization started from 1945 and culminated in 1973. The main features of this phase are given below.

(i) The mobility of capital was low than first phase.
(ii) The labour mobility was low.
(iii) The magnitude of free trade was low.
(iv) The global Institutions were getting created during this period.
(v) The National Institutions were heterogeneous and were not standardizes.

(c) Third Phase

The third phase of globalization started from 1974 and is still in progress. It is also known as the current phase of globalization. Main features of this phase are given below.

(i) The mobility of the capital is high.
(ii) The labour mobility is low.
(iii) The magnitude of free trade is extreme.
(iv) The International Institutes started functioning.
(v) The National Institutions have been standardized.
(vi) The on line global transactions are possible.
(vii) Information technology is being used extensively.

2. Types of Globalization

There are five main types of globalization, *viz.* (i) economic globalization, (ii) production globalization, (iii) cultural globalization, (iv) information globalization, and (v) ecological globalization. These are briefly discussed below:

(i) Economic Globalization

It refers to trans country flow of capital/finance or money. Main effects of economic globalization are:

(a) Enhancement in world wide economic relationships.
(b) Increase in international trade at a faster rate than the growth in the world economy.
(c) Increase in international flow of capital including foreign direct investment.
(d) Creation of international agreements leading to organizations like the WTO and OPEC.
(e) Development of global financial systems.

(f) Increased role of international organizations such as WTO, WIPO, IMF that deal with international transactions.

(g) Increase of economic practices like outsourcing, by multinational corporations.

(ii) Production Globalization

It refers to trans country flow of goods or products. It leads to trans national production of various goods or products. In such system, a product can be manufactured in several countries of the world with same quality.

(iii) Cultural Globalization

It refers to trans country flow of culture. The main effects of cultural globalization are:

(a) Greater international cultural exchange.

(b) Spreading of multiculturalism, and better individual access to cultural diversity.

(c) Greater international travel and tourism.

(d) Greater immigration, including illegal immigration.

(e) Spread of local foods such as pizza and Indian food to other countries.

(f) Development of a global telecommunications infrastructure and greater trans border data flow, using such technologies as the Internet, communication satellites and telephones.

(g) Increase in the number of standards applied globally; *e.g.* copyright laws and patents.

(h) Formation or development of a set of universal values.

(i) Spread of local goods, dresses and language to other countries.

(iv) Information Globalization

It refers to trans border flow of knowledge, ideas and information. It is also known as communication globalization or technological globalization. It makes use of information technology and permits on line global transactions.

(v) Ecological Globalization

It refers to global protection of ecosystem from degradation and pollution. Main features of ecological globalization are given below:

(a) It prevents ecosystem from various types of risks.

(b) It requires global collective action.

(c) It is also known as ecosystem globalization or environmental globalization.

(d) It leads to protection of environment.

3. Global Connectivity

Five types of global connectivity have been identified [Arjun Appadurai], *viz.* (i) ethnoscapes, (ii) financescapes, (iii) ideoscapes, (iv) media scapes, and (v) technoscapes. These are defined as follows:

Table 20.1: Types of Global Connectivity

Sl.No.	*Type of Global Connectivity*	*Refers to or Deals with*
1.	Ethnoscapes	Movements of people including tourists, immigrants, refugees and business travelers
2.	Financescapes	Global flow of capital or finance or money
3.	Ideo scapes	Global spread of ideas and political ideologies.
4.	Media scapes	Global distribution of media images that appear on our computer screen, in news papers, television and radios.
5.	Technoscapes	Global flow of technologies such green revolution in wheat and rice and information technology

4. Measurement of Globalization

The rate or extent of globalization is measured on yearly basis. Different types of globalizations are measured separately as follows:

Table 20.2: Measurement of the Rate of Globalization

Sl.No.	*Type of Globalization*	*Measured from*
1.	Economic Globalization	The data of trans border flow of capital or finance or money and direct foreign investment.
2.	Production Globalization	The data of trans country mobility of goods and products.
3.	Cultural Globalization	The mobility of tourists, travelers, and traders across the countries.
4.	Information Globalization	The data of information flow across the borders.
5.	Ecological Globalization	The work done for the protection of global ecosystem.

5. Factors Affecting Globalization

The rate of globalization is affected by several factors such as (i) global atmosphere, (ii) natural calamities, (iii) international relationships, (iv) means of transportation, (v) means of communications, (vi) tourists places, (vii) demand of goods, services, etc. These are briefly discussed below.

(1) **Global Atmosphere:** The peaceful global atmosphere promotes globalization, whereas the war situation restricts globalization. The globalization was adversely affected during First and Second World war period.

(2) **Natural Calamities:** Natural calamities such as earth quake, tornadoes, floods and disease epidemic have adverse effects on the rate of globalization.

(3) **International Relationships:** Harmonious relationships among countries enhance the rate of globalization, whereas disharmonious relationships restrict the process of globalization.

(4) **Means of Transportation:** Better means of transportation among countries promotes globalization, whereas poor transportation system restricts the globalization.

(5) **Means of Communication:** Better means of communication promotes globalization, whereas poor means of communication restricts the globalization.

(6) **Tourist Places:** Good and large number of tourist places in a country will attract tourists and travelers and vice versa.

(7) **Demand:** The demand of goods, services and information in other countries will enhance the international trade and the globalization.

6. Advantages of Globalization

Globalization has several advantages. Some important advantages of globalization are listed below.

1. **Connectivity:** People around the world are more connected to each other than ever before. Global mass media connects all the people in the world.
2. **Relationships:** The relationships between countries improve and the possibility of war between the developed countries decreases. The interdependence among Nations increases.
3. **Freedom of Trade:** It increases free trade between countries and reduces the international barriers.
4. **Investment opportunity:** As the liquidity of capital increases, developed countries can invest in developing ones The flexibility of corporations to operate across borders increases.
5. **Integration:** It leads to integration or consolidation of global markets. In other words the markets are interlinked. It is much easier for people to travel, communicate and do business internationally.
6. **Quality and Price:** There is improvement in quality and reduction in price due to competition among different companies.
7. **Flow of Goods and Services:** Information, money, technology and products flow across the border quicker than ever before. Products produced in one part of a country are available to the rest of the world. There is increase flow of communication between the individuals and corporations in the world. The movement of goods and people across the border is faster than ever before.
8. **Standard of Living:** Globalization offers a higher standard of living for people in rich countries and is the only realistic route out of poverty for the world's poor. It is claimed that globalization increases the economic prosperity and opportunity in the developing world. All the countries

involved in the free trade are at a profit. As a result, there are lower prices, more employment and a better standard of life in these developing nations.

9. **Efficient use of Resources:** The civil liberties are enhanced and there is a more efficient use of resources. The environmental protection in developed countries increases.
10. **Spread of Culture:** Globalization leads to better cultural understanding and tolerance. Due to improved transport facilities, more and more people are traveling to different countries, thereby spreading their culture to other parts of the world. Reduction of cultural barriers increases the global village effect There is spread of democratic ideals.

7. Disadvantages of Globalization

There are some disadvantages of globalization which are listed below.

1. **Increase in population:** Trans country flow of people will lead to increase in the population of certain countries especially in developed countries due to better facilities.
2. **Small Industries:** It will have adverse effects on small scale industries which can not compete in global market in terms of quality and price. Thus there will be hold of big industries.
3. **Employment:** There will be adverse effect on employment due to close down of small industries.
4. **Monetary Gain:** This will lead to tough competition among companies leading to loss in monetary gain.
5. **Terrorism:** Trans border flow of people may lead to increase in criminal activities and terrorism.
6. **Spread of Diseases:** There is greater risk of unintentional transmission of diseases between nations.
7. This may lead to widening of gap between rich and poor countries.
8. This may lead to exploitation of workers specially labours.

Effects of Globalization

The globalization has effects on movement of goods, services, information, finance, people, spread of cultures and ideas, markets, export and intellectual properties etc. These are briefly discussed below.

1. **Flow of Goods, Services Information etc.:** There is enhancement in the information flow between geographically remote locations and more trans border data flow using communication satellites, the Internet, wireless telephones etc.
2. **Markets:** The global common market has a freedom of exchange of goods and capital. The worldwide production and financial markets emerge. The free trade zones are formed having less or no tariffs.

3. **Access to Goods and Finance:** There is a broad access to a range of goods for consumers and companies. Corporate, national and sub-national borrowers have a better access to external finance.
4. **Solution of Global Problems:** Global environmental problems like cross-boundary pollution, over fishing on oceans, climate changes are solved by discussions International criminal courts and international justice movements are launched.
5. **Uniform Standards:** The standards applied globally like patents, copyright laws and world trade agreements increase.
6. **Spread of Culture:** Globalization leads to spread of cultures as there is individual access to cultural diversity. This diversity decreases due to hybridization or assimilation There is enhancement in worldwide fads and pop culture. The cross-cultural contacts grow and cultural diffusion takes place.
7. **Movement of People:** The International travel and tourism increases and immigration between countries increases The worldwide sporting events like the Olympic Games and the FIFA World Cup are held. The free circulation of people of different nations leads to social benefits.
8. **Export:** The local consumer products are exported to other countries. There is an increase in the desire to use foreign ideas and products, adopt new practices and technologies and be a part of world culture.
9. **Reduction in Transport Cost and Subsidies:** Due to development of containerization for ocean shipping, the transportation costs are reduced. The subsidies for local businesses decrease and there is reduction in capital controls.
10. **Recognition of Intellectual Property:** There is supranational recognition of intellectual property restrictions *i.e.* patents authorized by one country are recognized in another.

Summary

Globalization is the process of organizing the whole world into a single integrated marketing unit. It is also defined as the process of trans border free flow of products, services, people, culture, technology, and finance. It leads to the integration of economic, cultural, political, and social systems across national borders Globalization is also referred to as internationalization by some persons Both these terms are used as synonyms. However, some people use these terms separately. This chapter deals with various globalization such as phases of globalization, types [kinds] of globalization, global connectivity, measurement of globalization, factors affecting globalization, advantages of globalization and disadvantages of globalization.

Globalization is not a new phenomenon. It started with human civilization. In the past 130 years, modern historians have identified three stages or phases of globalization, *viz.*, first phase, second phase and third phase. These are briefly discussed.

There are five main types of globalization, *viz*. (i) economic globalization, (ii) production globalization, (iii) cultural globalization, (iv) information globalization, and (v) ecological globalization. These are briefly explained.

There are five types of global connectivity, *viz*. (i) ethnoscapes, (ii) financescapes, (iii) ideoscapes, (iv) media scapes, and (v) technoscapes. These have been defined. Methods of measuring globalization, factors affecting rate of globalization and advantages and disadvantages of globalization have been described.The globalization has effects on movement of goods, services, information, finance, people, spread of cultures and ideas, markets, export and intellectual properties etc. These have been briefly discussed.

Questions

1. **Define globalization and describe it advantages.**
2. **Discuss briefly disadvantages of globalization.**
3. **List different types of globalization. Describe any two of them in detail.**
4. **Explain briefly different phases of globalization.**
5. **Write short notes on the following:**
   (i) Production globalization (ii) Cultural globalization
   (iii) Economic globalization (iv) Ecosystem globalization
6. **Describe briefly different types of global connectivity.**
7. **Give a brief comparison of three phases of globalization.**
8. **What do the following terms signify:**
   (i) Finance scapes (ii) Idea scapes
   (iii) Media scapes (iv) Ethno scapes
9. **Describe in brief various factors affecting the rate of globalization.**
10. **Give a brief account of the methods of measuring globalization.**
11. **Discuss briefly the various effects of globalization.**
12. **Explain briefly the effects of economic and cultural globalization.**
13. **Give the full form of the following:**
   (i) IGO (ii) INGO
   (iii) NGO (iv) WIPO
24. **Define the following terms:**
   (i) Cosmopolitanism (ii) Cultural Imperialism
   (iii) Multiculturalism (iv) Global governance

Glossary

Accession: A sample of a crop variety collected at a specific location and time; may be of any size.

Act: The Essential Commodities Act, 1955.

Adaptation: The fitness of a genotype to a particular environment.

Additional Declaration: A statement that is required by an importing country to be entered in a phytosanitary certificate and which provides specific additional information pertinent to the phytosanitary condition of a consignment.

Alien species: Related species.

Allele: Alternative form of a gene.

Bio-control Agent: Any biological agent such as parasite, predator, parasitoid, microbial organism or self replicating entity that is used for control of pests.

Biodiversity: The totality of genes, species and ecosystem in a region or the world.

Bio-safety: Safe application of modern biotechnology with regard to human health, animal health and environment.

Breeders' Exemption: Rights provided to use protected material as the basis to develop a new variety or for other research work; also called research exemptions or breeders' privilege.

Candidate Variety: A variety that has to be protected.

Center of diversity: Geographic region with high levels of genetic or species diversity.

Competent Authority: An authority notified by the Central Government from time to time by notification in the Official Gazette.

Conservation of biodiversity: The management of biodiversity to meet requirement of the present and future generations.

Consignment: A quantity of seeds, plants and plant products or any regulated article consigned from one party to other at any one time shipment and covered by a phytosanitary certificate, bill of entry of customs, shipping/airway bill or invoice.

Container: A box, bottle, casket, tin, barrel, case, receptacle, sack, bag, wrapper or other thing in which any article or thing is placed or packed.

Controller: A person appointed as Controller of Seeds by the Central Government and includes any person empowered by the Central Government to exercise all or any functions of the Controller under this Order.

Copyrights: The exclusive rights to reproduce, sell and distribute a work, prepare derivative works and display the work publicly.

Cosmopolitanism: Attitude of concern for world as whole, interest in universal principles; commonly contrasted with nationalism.

Cotton: Includes ginned cotton, cotton linters and dropping, tripping, fly and other waste products of cotton mill other than yarn waste, but does not include cotton seed or un-ginned cotton.

Cultivar: A cultivated variety (genetic strain) of a domesticated crop plant.

Cultural Globalization: Trans country flow of culture.

Cultural Imperialism: Form of cultural hegemony enabling some states to impose worldview, values, and lifestyles on others.

Database Rights: A 'database right' is an intellectual property right given to a computer database.

Dealer: A person carrying on the business of selling, exporting or importing seeds, and includes an agent of a dealer.

Designated Inspection Authority: The authority notified by the Central Government from time to time through a notification to be published in the Official Gazette for the inspection of the plants grown in post entry quarantine facilities.

Distinctiveness: The variety must be distinguishable in at least one character from previously available varieties.

Domain Name: The names and words that companies designate for their registered Internet Web site addresses.

Domestic biodiversity: the genetic variation existing among the species, breeds, cultivars and individuals of animal, plant and microbial species that have been domesticated, often including their immediate wild relatives.

Ecological Globalization: Global protection of ecosystem from degradation and pollution.

Entry Point: Sea port, airport or land customs stations through which import is permitted under this order.

Essentially derived variety: A variety derived from the initial variety and having most of the characteristics of initial variety but clearly distinguishable from initial variety.

Ethnoscapes: Movements of people including tourists, immigrants, refugees and business travelers.

Example variety: A variety that is used for comparison of a particular character.

Exclusive Rights: The legal privileges that only a copryight owner has with respect to their copyrighted work.

Export: Taking out of India by land, sea or air.

Export: To take or cause to be taken out from any place in India to a place outside India.

Export-processing zones: Also free trade zones. Selected areas in industrializing countries marked by low taxes and tariffs, subsidized infrastructure, and exemption from some regulations, designed to attract foreign direct investment and stimulate growth.

***Ex situ* conservation:** A conservation of biodiversity alive outside of its original habitat or natural environment.

Extant variety: A variety available in India such as farmers' variety, variety of common knowledge or any other variety of public domain.

Extinction: The permanent loss of a species caused due to natural failure to adapt to environmental change.

Farmer: Any person who cultivates crops either by cultivating the land himself or through any other person but does not include any individual, company, trader or dealer who engages in the procurement and sale of seeds on a commercial scale.

Farmers' Exemption: Legal rights provided to farmers to save, use, exchange, share or sell his farm produce of a protected variety; also called farmers' rights or farmers' privilege.

Farmers' Rights: The legal rights provided to farmers to save, use, sow, replant, exchange, share or sell his farm produce including seed of a variety protected under Plant Variety Protection Act.

Farmers' Variety: A variety that has been developed by a farmer and used for commercial cultivation for several years.

Finance scapes: Global flow of money.

Form: A form appended to this Order.

Fruit: Any fleshy portion of the plant, that contains seeds, which is used for consumption, including seedless fruit both fresh and dry but does not include preserved or prickled or frozen fruits.

Gene bank: Various organizations where genetic diversity is maintained in living state.

Gene: The functional unit of heredity; the part of the DNA molecule that encodes a single enzyme or structural protein unit.

Genetic diversity: Variety of genes and genotypes found in a particular crop species.

Genetic Resources: The genetic material of actual or potential value.

Genotype: The set of genes possessed by an individual organism.

Geographical Indication: A geographical indication is a sign used on goods that have a specific geographical origin and often possess qualities or a reputation that are due to that place of origin.

Germplasm: Plants in whole or in parts and their propagules including seeds, vegetative parts, tissue cultures, cell cultures, genes and DNA based sequences that are held in a repository or collected from wild as the case may be and are utilized in genetic studies or plant breeding programs for crop improvement.

Global governance: Rules and institutions for managing and regulating actions or processes of global import.

Globalization: The process of trans border free flow of products, services, people, culture, technology, and finance.

Grain: Seeds intended for processing or consumption and not for sowing or propagation.

Horticulture nursery: Any place where horticulture plants are, in the regular course of business, produced or propagated and sold for transplantation.

Ideoscapes: Global spread of ideas and political ideologies.

IGO (Intergovernmental organization): Formed by and membership restricted to states. Examples: UN, NATO.

Immovable Property: Fixed types of properties which cannot move from one place to other. Such properties include land, buildings [house, flat, bungalow, farm house] and gardens.

Import Permit: An official document authorizing importation of a consignment in accordance with specified phytosanitary requirements.

Import: An act of bringing into any part or place of territory of Republic of India any kind of seed, plant or plant product and other regulated article from a place out side India either by sea, land, air or across any customs frontier.

Import: To bring or cause to be brought to any place in India from outside India.

Indigenous Peoples: Groups held to be original residents of certain areas, especially illiterate groups under threat of displacement due to development.

Industrial Design: Any original shape, picture, or some combination applied to a useful article of manufacture. In other words, it refers tothe design of the mass-produced products of our everyday environment, from sinks and furniture to computers.

Information Globalization: Trans border flow of knowledge, idea and information.

INGO: International nongovernmental organization.

***In situ* conditions**: The conditions where genetic resources exist within ecosystems and natural habitats.

Inspection Authority: An authority notified by the Central Government from time to time or an officer of the Directorate of Plant Protection, Quarantine and Storage duly authorized by the Plant Protection Adviser (PPA) for the purpose of approval and certification of post-entry quarantine facilities and inspection of growing plants in such facilities in accordance with guidelines issued by PPA.

Inspector: An inspector of seeds appointed under Clause 12.

Intellectual Property Rights: The legal rights provided to an inventor to derive economic benefits from his invention/innovation.

Intellectual Property: The product/process/idea which is outcome of the brain of a person and can be used on commercial scale for benefit of human kind.

Issuing Authority: An authority as envisaged under Schedule-IV of this order or duly notified by the Central Government from time to time either generally or specifically for issuance of import permit.

Keystone species: A species whose loss from an ecosystem would cause a greater than average change in other species populations or ecosystem processes.

Kind: One or more related species or sub-species of crop plants each individually or collectively known by one common name such as cabbage, maize, paddy and wheat.

Landrace: Primitive crop cultivars which were selected and cultivated by farmers for many generations.

Landraces: A crop cultivar or animal breed that evolved with and has been genetically improved by traditional agriculturalists, but has not been influenced by modern breeding practices.

Mask Work: A mask work is a two or three-dimensional layout of an integrated circuit.

Mediascapes: Global distribution of media images that appear on our computer screen, in news papers, television and radios.

Misbranded: A seed with false label or label of another variety.

Moral Rights: Moral rights are a special extension to copyright that gives the originator of a copyrighted work rights over its use.

Movable Property: The property which can be easily shifted from one place to other. Such property includes animals, farm machines, furnitures, fixtures, gold, silver, diamond, and money.

Multiculturalism: Doctrine asserting value of different cultures coexisting within single society; globally, vision of cultural diversity deliberately fostered and protected.

NGO: Nongovernmental organization.

Notification: A notification published in the official Gazette and the expression "notifies" shall be construed accordingly.

Novelty: Newness of a variety. A variety which has not been grown for more than one year prior the application for registration.

Noxious Weeds: Any weed harmful or hazardous or unwholesome to human beings, animal life or parasitic on plant species.

Nursery: Any orchard, or any other place, facility, glass-house, screen house, utilized for raising plants.

Official Phytosanitary Certificate: A phytosanitary certificate in the format (reproduced as Schedule I) prescribed by the International Plant Protection Convention sponsored by the Food and Agricultural Organization of the United Nations Organization and issued by the authorized officer of the country of origin of consignments.

Orthodox seed: Seed that can be dried to moisture levels between 4 and 6 percent and kept at low temperatures.

Packing Material: Any kind of material of plant origin used for packing, which shall include hay, straw, wood savings, wood chips, saw dust, wood waste, wooden pallets, dunnage mats, wooden packages, coir pith, peat or sphagnum moss etc.

Patent: A document granting an inventor sole rights to an invention. It is an official document which grants sole rights to the inventor for manufacturing and marketing his product/process/invention to derive benefits.

Pest Risk Analysis: The process of evaluating biological or other scientific and economic evidence to determine whether a pest should be regulated and strength of any phytosanitary measures to be taken against it.

Pest: Any biotic agent capable of causing any injury or damage to plants and plant products and include any form or stage of insects, mites, snails, slugs, worms, nematodes, algae, fungi, protozoa, bacteria, actinomycetes, viruses, viroids and molecules and also include genetically engineered or modified organisms and weeds.

Phenotype: The external appearance of an individual.

Phytosanitary Certificate: A certificate issued in the model format prescribed under the International Plant Protection Convention of the Food and Agricultural Organization and issued by an authorized officer at the country of origin of consignment or re-export.

Plant Breeders' Rights: Plant breeders' rights [PBR], also known as plant variety rights (PVR), are intellectual property rights granted to the breeder of a new variety of plant.

Plant Product: An un-manufactured material of plant origin including grain and those manufactured products that, by their nature or that of their processing, may create risk for the introduction and spread of a pest.

Plant Protection Adviser: The plant Protection Adviser to the Government of India, Directorate of Plant Protection, Quarantine and Storage, N.H.IV, Faridabad.

Plant: Any plant or part thereof, whether living or dead, trees, shurbs, nursery stock, and includes all vegetatively propagated materials.

Point of Entry: Any sea port, airport, or land-border check-post or rail station, river port, foreign post office, courier terminal, container freight station or inland container depot notified as specified in Schedule-I or Schedule-II or Schedule-III as the case may be.

Post Entry Quarantine: Growing of imported plants in confinement for a specified period of time in a glass house, screen house, poly house or any other facility, or isolated field or an off-shore island that is established in accordance with guidelines/standards and are duly approved and certified by an inspection authority notified under this order.

Producer: A person, group of persons, firm or organization who grow or organize the production of seeds.

Production globalization: Trans country flow of goods or products.

Property: The wealth or valuable things earned by a person.

Public Domain: The knowledge that is freely available, commonly shared throughout the world without any access restrictions. It cannot be protected by intellectual property rights.

Quarantine Pest: A pest of potential economic importance to the area endangered thereby and not yet present there, or present but not widely distributed and being officially controlled.

Recalcitrant seed: Seed that does not survive drying and freezing.

Reference variety: All released and notified extant varieties that are under seed production chain.

Registering Authority: A licensing authority appointed under clause 11.

Regulated Article: Any article the import of which is regulated by this order.

Schedule: Schedule annexed to this order.

Seed bank: A facility designed for the *ex situ* conservation of individual plan varieties through seed preservation and storage.

Seed processing: The process by which seeds and planting materials are dried, threshed, shelled, ginned or delinted (in cotton), cleaned, graded or treated.

Seeds: Seeds of agricultural and horticultural crops and forest plant species produced by sexual reproduction and shall include naked seeds (cones) produced by gymnosperms and seed sprouts meant for propagation or consumption.

Selection: The process that favors survival and further propagation of some plants having more desirable characters than others. Natural selection favors those traits which are essential for survival of a species. Artificial selection favors those characters which are of economic value to the human.

Soil: Earth, sand, clay, silt, loam, compost, manure, peat or sphagnum moss, litter, leaf waste or any organic media that support plant life and shall include ship ballast or any organic medium used for growing plants.

Species diversity: Variety of species found in a given region or area.

Species: A group of organisms capable of interbreeding freely with each other but not with members of other species.

Spurious seed: Any seed which is not genuine or true to type.

Stability: Each generation of the variety should be similar.

State Government in relation to a Union Territory: The Administrator hereto by whatever designation known.

Subspecies: A subdivision of a species.

***Sui Generis* Rights:** *Sui-generis* refers to things of their own kind or things with unique characteristics.

Supplementary Protection Certificate: This term is used in European Countries. This is an extension of the term given to pharmaceutical or plant protection patent.

Sustainable Development: Policy of promoting growth consistent with protection of environment.

Technoscapes: Movement of technologies around the world.

Timber: A form of dead wood, log and lumber cut from plants, with or without bark or sawn and sized, which is used for manufacturing veneer, plywood, particle or chip board and making building material, furniture, packages, pallets, sports goods and handicrafts.

Tissue Cultured Plant: Any part of a plant or plant tissue or plantlet grown under aseptic or sterile conditions in flasks or other suitable container on appropriate media and shall include ex-agar washed plant lets.

Trade Name: The name under which a company conducts business, or by which its business, product/service are identified. It may or may not be registered as a trademark.

Trade Secret: A formula or process or device used in business that is not published or divulged which gives an advantage over competitors.

Trademark: A trademark can be a word, name, symbol, device or mark which is used to identify and distinguish the goods or services of one company from goods or services of another.

Traditional Knowledge: The Traditional Knowledge [TK] refers to the knowledge, innovations and practices of indigenous people and local communities.

Transgenic Variety: Seed or planting material synthesized or developed by modifying or altering the genetic composition by means of genetic engineering.

Uniformity: All plants in a variety should look alike.

Variety: A plant grouping within a single botanical *taxon* of the lowest known rank.

WTO: An international organization that regulates trade and tariffs between nations in order to ensure that trade flows smoothly, predictable and as freely as possible.

References

A. SEED LEGISLATIONS

1. Anonymous, 1966. Indian Seed Act 1966 [Act No. 54 of 1966]. Government of India Ministry of Agriculture (Department Of Agriculture), December, 29, 1966.
2. Anonymous, 1968. Seed Rules, 1968. Government of India Ministry of Agriculture (Department of Agriculture), 1968.
3. Anonymous, 1973. The Seeds (Amendment) Rules, 1973, Government of India Ministry of Agriculture (Department of Agriculture), No. 7(17)/69-Seeds Dev. New Delhi, Dated The 30.6.1973.
4. Anonymous, 1983. The Seeds (Control) Order, 1983. Under seed Act, 1966 (Act No. 54 Of 1966). Government Of India, Ministry of Agriculture, (Department of Agriculture and Cooperation), New Delhi, Dated the 30th Dec. 1983.
5. Anonymous 1988. New Seed Policy. Ministry of Agriculture, vide Letter No.11-71/88-SD-1 dated September 16,1988.
6. Anonymous, 1989. The Plants, Fruits and Seeds (Regulation of Import into India) Order, 1989 Ministry of Agriculture (Department of Agriculture and Cooperation), New Delhi, the 27th October, 1989.
7. Anonymous, 2001. The Protection of Plant Varieties and Farmers' Rights Act, 2001, Ministry of Law, Justice and Company Affairs (Legislative Department), *New Delhi, the 30th October, 2001.*
8. Anonymous, 2002. National Seed Policy, 2002. Government of India Ministry of Agriculture (Department of Agriculture), 19th June, 2002.
9. Anonymous, 2002. The Biological Diversity Act, 2002, Government of India, 2002.

10. Anonymous, 2004. Seed Bill 2004. Government of India Ministry of Agriculture (Department of Agriculture), March, 2004.

11. UNIVERSAL, 2006. The protection of Plant Varieties and Farmers' Rights Act, 2001 and The protection of Plant Varieties and Farmers' Rights Rules, 2003. Bare Act with Short Notes. Universal Law Publishing Co. Pvt. Ltd., New Delhi.

12. Singh Phundan 2013. Essentials of Plant Breeding, 5th edn. Kalyani Publishers, New Delhi.

13. Singh Phundan, 2009. IPR and Plant Breeders' Rights [Subjective]. New Vishal Publications, New Delhi.

14. Singh Phundan, 2010. IPR and Plant Breeders' Rights [Objective]. New Vishal Publications, New Delhi.

15. Singh Phundan, 2010. IPR and Plant Breeders' Rights [At a Glance]. New Vishal Publications, New Delhi.

B. PLANT BREEDING

1. ALLARD, R. W. 1960. Principles of Plant Breeding. John Wiley and Sons Inc. New York.

2. ALLARD, R. W. 1999. Principles of Plant Breeding 2nd edn. John Wiley and Sons Inc. New York.

3. BOROJEVIC, S. 1990. Principles and Methods of Plant Breeding. Alsevier, Amsterdam.

4. BRIGGS, F. N. AND KNOWLES, P. F. 1967. Introduction to Plant Breeding. Reinhold, New York.

5. CHOPRA, V. L. (ED.) 2000. Plant Breeding: Theory and Practice 2nd edn. Oxford and IBH Publishing Company, Pvt. Ltd. New Delhi.

6. FEHR, W. R. !987. Principles of Cultivar Development Volume 1. Theory and Techniques. Macmillan Publishing Company, New York.

7. FREY, K. J. (ed.) 1966. Plant Breeding. Iowa State University, Press, Ames, Iowa, USA.

8. FREY, K. J. (ed.) 1981. Plant Breeding II. Iowa State University, Press, Ames, Iowa, USA.

9. HARTEN, A. V. 1998. Mutation Breeding: Theory and Practical Applications. Cambridge Uni. UK.

10. KUCKKUCK, H., KOBADE, G. AND WENGEL G. 1993. Fundamentals of Plant Breeding. Narosa Publishing House, New Delhi.

11. MAYO, O. 1984. The Theory of Plant Breeding. Clarendon, Oxford.

12. POEHLMAN, J. M. 1987. Breeding Field Crops 3rd edn.AVI Publishing Co. Inc.West Port, Connecticut, USA.

13. POEHLMAN, J. M. AND BORTHAKUR, D. N. 1969. Breeding Asian Field Crops. Oxford and IBH Publishing Company, New Delhi.

14. PHUNDAN SINGH 2030. Essentials of Plant Breeding 5th edn. Kalyani Publishers, New Delhi.
15. PHUNDAN SINGH 2009. Genetics 2nd edn. Kalyani Publishers, New Delhi.
16. PHUNDAN SINGH 2012. Cotton Breeding 3rd edn. Kalyani Publishers, New Delhi.
17. PHUNDAN SINGH AND SANJEEV SINGH 2010. Breeding Hybrid Cotton 3rd edn. Kalyani Publishers, New Delhi.
18. SHARMA, J. R. 1984. Principles and Practice of Plant Breeding. Tata McGraw-Hill Publishing Company Ltd. New Delhi.
19. SIMMONDS, N. W. 1979. Principles of Crop Improvement. Longman, London.
20. SNEEP, J. AND HENDRIKSON, A. J. T. (eds) 1979. Plant Breeding Perspectives. Pudoc, Wageningen, Netherlands.
21. VOSE, P. B. AND BLIXT, S. G. (eds) 1984. Crop Breeding A Contemporary Basis. Pergamon Press, London.
22. WILLIAMS, W. 1964. Genetic principles and Plant Breeding, Oxford, Blackwell.

C. GENETICS

1. ACQUAAH GEORGE. 2006. Principles of Plant Genetics and Breeding. Blackwell Publishers.
2. CARROLL, M. 1989. Organelles. Macmillan Press Ltd. Basing Stock, U.K.
3. DOUCE, R. 1985. Mitochondria in Higher Plants. Academic Press, Inc. Orlando, Florida, USA.
4. FRIEFELDER, D. 1987. Molecular Biology 2nd ed. Jones Bartlett Publishers, Inc. USA.
5. GARDNER, E. J., SIMMONS, M. J. AND SNUSTED, D. P. 1991. Principles of Genetics 8th ed. John Wiley and Sons Inc., New York, USA.
6. HERSKOWITZ, I. H. 1979. Elements of Genetics. Macmillan Publishing Co. Inc., New York, USA.
7. KHUSH, G. S. 1976. Genetics of Aneuploids 2nd ed. Academic Press Inc., New York, USA.
8. SNUSTED, D. P. AND SIMMONS, M. J. 2005. Principles of Genetics. John Wiley and Sons, Inc., New York, USA.
9. PHUNDAN SINGH 2009. Genetics 2nd edn. Kalyani Publishers, New Delhi.
10. PHUNDAN SINGH 2011. Elements of Genetics 3rd ed. Kalyani Publishers, New Delhi.
11. STRICKBERGER, M. W. 1985. Genetics. 3rd ed. Macmillan Publishing Co., New York, USA.
12. SINGH, B. D. 1995. Fundamentals of Genetics. 2nd ed. Kalyani Publishers, New Delhi.

C. STATISTICS

1. CHANDEL, S.R. S. 1993. A Hand Book of Agricultural Statistics, 3rd ed. Achal Prakashan Mandir, Kanpur, India.
2. DAS, M. N. AND GIRI, N. C. 1986. Design and Analysis of Experiments, 2nd ed. Wiley Eastern Limited, New Delhi.
3. ELHANCE, D. N. AND ELHANCE, V. 1990. Fundamentals of Statistics. Kitab Mahal, New Delhi.
4. GUPTA, S. K. AND KAPOOR, V. K. 1993. Fundamentals of Mathematical Statistics. S. Chand and Sons, New Delhi.
5. RAO, C. R. 1952. Advanced Statistical Methods in Biometrical Research. Wiley and Sons, New Delhi.
6. SINGH, S., SINGH, T. P., BANSAL, M. L. AND KUMAR, R. 1991. Statistical Methods for Research Workers. Kalyani Publishers, New Delhi.
7. PANSE, V. G. AND SUKHATME, P. V. 1985. Statistical Methods for Agricultural Workers. ICAR Publication, New Delhi.
8. FISHER, R. A. 1970. Statistical Methods for Research Workers 14th ed. Oliver and Boyd, London.

D. BIOMETRICAL GENETICS

1. FALCONER, D. S. 1989. Introduction to Quantitative Genetics, 3rd ed. Longman, New York.
2. HALLAUER, A. R. AND MIRANDA, J. B. 1981. Quantitative Genetics in Maize Breeding. Iowa State Uni. Press Ames, USA.
3. KEMPTHORNE, O. 1957. An Introduction to Genetic Statistics. John Wiley and Sons Inc., New York.
4. MATHER, K. 1949. Biometrical Genetics. Methuem and Co. Ltd. London.
5. MATHER, K. AND JINKS, J. L. 1971. Biometrical Genetics. Chapman and Hall, London.
6. SINGH, PHUNDAN AND NARAYANAN, S. S. 2013. Biometrical Techniques in Plant Breeding, 4rd ed. Kalyani Publishers, New Delhi.
7. SINGH, R. K. AND CHAUDHARY, B. D. 1985. Biometrical Methods in Quantitative Genetic Analysis, 2nd ed. Kalyani Publishers, New Delhi.

E. SEED TECHNOLOGY

1. AGRAWAL, P. K. 1994. Principles of Seed Technology. ICAR, New Delhi.
2. AGRAWAL, P. K. AND DADLANI, M. (eds) 1987.Techniques in Seed Science and Technology. South Asian Publishers, New Delhi.
3. AGRAWAL, R. L. 1980. Seed Technology. Oxford and IBH Publishing Co. New Delhi.

4. BOYCE, K. 1981. Introduction to Seed Science and Technology. ICARDA, Aleppo, Syria.
5. CHALAM, G. V., SINGH, A. AND DOUGALS, S. J. F. 1967. Seed Testing Manual. ICAR and USAID, New Delhi.
6. COPELAND, L. O. AND MCDONALD, M. D. 1985. Principles of Seed Science and Technology. Burgess Publishing Company, Minneopolis, Minnesota, USA.
7. DAHIYA, B. S. AND RAI, K. N. (eds) 1997. Seed Technology. Kalyani Publishers, New Delhi.
8. DOUGHLASD, J. E. 1976. Seed Certification Manual. National Seeds Corporation and Rockefeller Foundation, New Delhi, India.
9. JOSHI, A. K. AND SINGH, B. D. 2004. Seed Science and Technology. Kalyani Publishers, New Delhi.
10. KHARE, D. AND BHALE, M. S. 2000. Seed Technology. Scientific Publishers, Jodhpur.
11. NSC and RF, 1972. Field Inspection Manual. NSC and Rockefeller Foundation. New Delhi.
12. NSC, 1972. A Hand Book for Seed Inspectors. National Seeds Corporation, New Delhi.
13. SINGH PHUNDAN 2010. Essentials of Plant Breeding. Kalyani Publishers, New Delhi.
14. ROBERTS, E. H. 1972. Viability of Seeds. Chapman and Hall, London.

F. PLANT BIOTECHNOLOGY

1. BAJAJ, Y.P.S. 1994. Somatic Hybridization in Crop Improvement. Springer Verlag.
2. BHOJWANI, S. S. (ed.) 1990. Plant Tissue Culture: Applications and Limitations. Elsevier, Amsterdam.
3. BHOJWANI, S. S. AND BHATNAGAR, S. P. 1990. Embryology of Angiosperms. Vikas Publishing Co., New Delhi.
4. BHOJWANI, S. S. AND RAJDAN, M. K. 1996. Plant Tissue Culture: Theory and Practice. Revised Edition. Elsevier, Amsterdam.
5. CHAWLA, H. S. 2002. Introduction to Plant Biotechnology. 2nd ed. Oxford and IBH Publishing Co.PVT. LTD., New Delhi.
6. KHANNA, V. K. 1999. Plant Tissue Culture. Kalyani Publishers, New Delhi.
7. PHUNDAN SINGH 2006. Introduction to Biotechnology. Kalyani Publishers, New Delhi.
8. PHUNDAN SINGH 2012. Plant Biotechnology. Kalyani Publishers, New Delhi.
9. PRAKASH, J AND PIERIK, R.L.M. (eds.). 1993. Plant Biotechnology: Commercial Prospects and Problems. Oxford and IBH Publishing Co.PVT. LTD., New Delhi.

10. RAJDAN, M. K. 1993. An Introduction to Plant Tissue Culture. Oxford and IBH Publishing Co.PVT. LTD., New Delhi.

11. Sensen, C. W. (ed.) 2002. Essentials of Genomics and Bioinformatics. Wiley-VCH.

12. SINGH, B. D. 2006. Plant Biotechnology. Kalyani Publishers, New Delhi.

13. VASIL, I. K. AND THORPE T. A. 1994. Plant Cell and Tissue Culture. Kluwer Academic Publishers, Dordrecht, Netherlands.

14. ZAITLIN, M., DAY, P. AND HOLLAENDER, A. 1985. Biotechnology in Plant Science: Relevance to Agriculture in the Eighties. Academic Press, N.Y.

Books by the Same Author

A. Plant Breeding

1. Essentials of Plant Breeding
2. Plant Breeding: Molecular and New Approaches
3. Principles of Seed Technology
4. Practical in Crop Breeding
5. Practical and Numerical in Plant Breeding
6. Numerical Problems in Plant Breeding and Genetics
7. Objective Science of Plant Breeding
8. Plant Breeding At a Glance
9. Plant Breeding [For Under Graduate Students]
10. Principles of Plant Breeding
11. Cotton Breeding
12. Heterosis Breeding in Cotton
13. Breeding Hybrid Cotton
14. Breeding Transgenic Bt. Cotton
15. Cotton Improvement in India
16. Glossary of Plant Breeding and Genetics
17. IPR and Plant Breeders' Rights [Subjective]
18. IPR and Plant Breeders' Rights [Objective]
19. IPR and Plant Breeders' Rights [At a Glance]

20. Elements of Baby Corn
21. Breeding Crop Plants for Stress Resistance
22. Plant Breeding [Related Legislations]

B. Genetics

23. Elements of Genetics
24. Genetics
25. Principles of Genetics [For UG Students]
26. Plant Genetics
27. Cotton Genetics
28. Objective Genetics
29. Genetics At a Glance
30. Objective Genetics and Plant Breeding
31. Objective Genetics and Plant Breeding [Hindi Edn]
32. Glossary cum Dictionary of Genetics and Plant Breeding
33. Molecular Genetics [Objective]
34. Molecular genetics [At a Glance]
35. Molecular Genetics [Subjective]
36. Genetics and Man
37. Genetics Today

C. Quantitative Genetics

38. Biometrical Techniques in Plant Breeding
39. Application of Biometrical Techniques in Plant Breeding [Hindi Edn]
40. Quantitative Genetics
41. Objective Quantitative Genetics
42. Quantitative Genetics At a Glance

D. Biotechnology

43. Introduction to Biotechnology
44. Plant Biotechnology
45. Principles of Plant Biotechnology

E. Miscellaneous

46. A Guide for Competitive Examinations of Agriculture

www.ingramcontent.com/pod-product-compliance
Ingram Content Group UK Ltd.
Pitfield, Milton Keynes, MK11 3LW, UK
UKHW041644031026
15587UKWH00001B/477

9 789351 308546